HYDROPONICS

The Ultimate Guide to Grow your own Hydroponic Garden at Home: Fruit, Vegetable, Herbs.

LARA DARLING

Copyright © 2019 Lara Darling
All rights reserved.
ISBN: 9781696731423

TABLE OF CONTENTS

Introduction 16
Part 1 - HYDROPONICS 19
Chapter 1 - BASIC CONCEPTS . 19
    Wick Systems............................. 20
    Water Culture 21
    Ebb and Flow............................. 22
    Choosing What to Grow 22
    Lighting 23
    Room Conditions....................... 24
    Water Quality 25
    Nutrients 26
    Additional Equipment............... 27
    Good Starter Plants 28
Chapter 2 - WHY HYDROPONICS 31
    1. Space Savings 31
    2. Hydroponics Saves Water..... 32
    3. No Weeding Necessary......... 33

4. Less Pests and Diseases - Hydroponic Pests........................ 34

5. Double-Headed Time Savings 34

6. Gives You EXTREME Control 35

Chapter 3 - WHAT HYDROPONICS IS 38

Pros of hydroton............................. 44

1) High pore space means fewer blockages...................................... 44

2) Some air-holding capability to hold root zones oxygenated...... 44

3) Fairly renewable & environment-friendly.................. 45

4) Reusable.................................. 45

5) Easy to plant and harvest 46

6) Good colonization for microbial populations................ 46

Cons of hydroton 47

1) Water maintaining potential leaves something to be desired. 47

2) Fairly costly 47

3) Can reason troubles with pumps and plumbing.................. 48

Pros of perlite................................. 48

What is perlite?............................ 48

Pros of perlite.............................. 49

1) Perlite is usually reusable........ 49

2) Perlite helps deal with anaerobic conditions. 49

3) Perlite is inexpensive............. 50

4) Sterile and pH neutral........... 50

Cons of perlite................................ 51

1) Perlite ore is not a renewable resource. 51

2) Aggressive root structures can motive blockage. 52

3) Vulnerable to solids loading. 52

4) Mud hurts fish and may be dangerous if inhaled.................... 53

How to Use Rockwool as Medium and also the edges It Offers...... 55

Advantages Of Rockwool.......... 56

How To Use Rockwool 57

What Is Coco Coir And How Is It Made? 60

Basic Types Of Coco Coir 61

The Benefits Of Using Coconut Coir 62

What Are The Drawbacks Of Using Coco Coir? 64

Coco Coir Features That May Be A Pro Or A Con 65

What Are Hydroponic Hydrogels? 66

What Makes Hydroponic Hydrogels so Useful? 67

Benefits of Hydroponic Hydrogels 68

Chapter 4 - HYDROPONICS SYSTEMS 70

Types of Hydroponic Systems . 70

Benefits of Wick Systems 71

Benefits of Deep Water Culture 73

Nutrient Film Technique (NFT) Systems 73

Benefits of Nutrient Film Technique 74

Benefits of Ebb and Flow 76

Benefits of Aeroponics 77

Benefits of Drip Systems 78

Chapter 5 - HOW TO CHOOSE THE RIGHT HYDROPONIC SYSTEM 80

1. Available space 80

2. Automation 81

3. Expandability of the Hydroponics System 82

4. Energy efficiency 82

5. System charge and setup costs 83

Chapter 6 - ADVANTAGES AND DISADVANTAGES OF HYDROPONICS 85

1. No soils needed 85

2. Make better use of house and location 85

3. Climate control 86

4. Hydroponics is water-saving 86

5. Effective use of nutrients 87

6. pH manage of the solution ... 88

7. Better increase rate 88

8. No weeds 88

9. Fewer pests & diseases . 89

10. Less use of insecticide, and herbicides 89

11. Labor and time savers 90

12. Hydroponics is a stress-relieving hobby 90

13. An aquicultural garden needs it slow and commitment 91

14. Experiences and technical knowledge 92

15. Organic debates 92

16. Water and electrical energy risks 93

17. System failure threats 93

18. Initial expenses 94

19. Long return per investment 94

20. Diseases and pests may additionally unfold quickly 95

Chapter 7 - SUITABLE CROPS IN HYDROPONICS 96

TOP 5 PLANTS FOR NEW HYDROPONIC GARDENS . 96

LETTUCE IN HYDROPONICS 96

SPINACH IN HYDROPONICS 97

STRAWBERRIES IN HYDROPONICS 98

BELL PEPPERS IN HYDROPONICS 99

HERBS IN HYDROPONICS 100

Part 2 - YOUR OWN HYDROPONICS 101

Chapter 8 - HOW TO START YOU OWN HYDROPONIC GARDEN 101

Preparation of the Grow Media 103

Determining the Correct Watering Cycle for Your Plants 105

Planting and Growing Tips for Hydroponics 107

Nutrition Tips for Hydroponics 109

Chapter 9 - HOW TO SELECT THE MOST APPROPRIATE GROWING MEDIUM............... 110

Choosing the Right Medium For Your Hydroponic System 110

Rockwool 110

Coconut Fiber 111

Perlite... 111

Expanded Clay Pebbles........... 112

Air 112

Chapter 10 - CHOOSING PLANTS 114

What Can You Grow Hydroponically? 114

Flowers 114

Herbs 114

Anise... 115

Basil 115

- Cannabis 116
- Rosemary 116
- Tarragon 117
- Thyme 118
- Watercress 118
- Vegetables 119

Chapter 11 - TRANSPLANTING 120

- How to Clone Hydroponic Plants and Transplant 120

Chapter 12 - HOW TO SET UP YOUR OWN HYPROPONIC GARDEN 124

1. The Passive Bucket Kratky Method 125
2. Simple Bucket Hydroponic System 125
3. Simple Drip System With Buckets 126
4. Aquarium Hydroponics Raft 127
5. PVC NFT Hydroponics System 127

Chapter 13 - PROBLEMS AND TROUBLES 129

  1. Hydroponics System Leaks 129

  2. Buying Cheap, Insufficient Or Incorrect Lighting 131

  3. Using The Wrong Fertilizer 132

  4. Not Keeping Things Clean. 133

  5. Not Learning As You Go ... 134

  6. Not Monitoring The Health Of Your Plants 135

  7. Not Monitoring And Adjusting the pH Level 135

  8. Nutrient Deficiency and Toxicity 137

Chapter 14 - COMMON MISTAKE TO AVOID 139

  Mistake #1: Growers layout unusable or hard-to-use farms 139

  Mistake #2: Growers underestimate manufacturing and device costs 140

Mistake #3: Producers confuse biological viability with monetary viability 142

Mistake #4: Growers pick out the wrong plants for their local weather or technique 143

Mistake #5: Growers choose the incorrect market 144

Mistake #6: Growers operate structures that have terrible tune records, then count on distinctive results 145

Mistake #7: Growers grow too big, too fast 146

Hydroponics Gardening 149

Chapter 15 - TIPS AND TRICKS 151

Conclusion 153

Introduction

Look, no soil! We're so used to plants developing in fields and gardens that we locate anything else completely extraordinary. But it is true. Not only will plant life develop without soil, they often develop a lot better with their roots in water or very moist air instead. Growing flora besides soil is known as hydroponics. It may sound weird, but many of the meals we eat—including tomatoes on the vine—are already grown hydroponically. Let's take a nearer seem at hydroponics and find out how it works!

Many one of a kind civilizations have utilized hydroponic developing methods all through history. As stated in Hydroponic Food Production by Howard M. Resh: "The placing gardens of city, the floating gardens of the Aztecs of Mexico and those of the Chinese are examples of 'Hydroponic' culture. Egyptian hieroglyphic records dating returned several hundred years B.C. describe the growing of flora in water." While hydroponics is an ancient approach of growing plants, giant strides have been made over the years in this modern region of agriculture.

Throughout the ultimate century, scientists and horticulturists experimented with one-of-a-kind methods of hydroponics. One of the attainable applications of hydroponics that drove lookup was developing fresh produce in non-arable areas of the world and areas with little to no soil. Hydroponics was once used throughout war II to grant troops stationed on non-arable islands within the Pacific with clean turn out fully grown in domestically established aquacultural systems.

Later in the century, hydroponics was once built-in into the space program. As NASA regarded the practicalities of finding a society on any other planet or the Earth's moon, hydroponics easily in shape into their sustainability plans. By the 1970s, it wasn't just scientists and analysts who were concerned with hydroponics. Traditional farmers and eager hobbyists commenced to be attracted to the virtues of hydroponic growing.

A few of the benefits of hydroponics include:

The potential to produce greater yields than traditional, soil-based agriculture.

Allowing meals to be grown and fed on in areas of the world that can't support plants in the soil.

Eliminating the need for huge pesticide use (considering most pests stay in the soil), successfully making our air, water, soil, and food cleaner.

Commercial growers are flocking to hydroponics like in no way before. The beliefs surrounding these developing techniques contact on subjects that most humans care about, such as supporting give up world starvation and making the world cleaner. People from all over the world have been building or purchasing their systems to develop great-tasting, sparkling food for their family and friends. Ambitious folks are striving to make their desires come true by way of making their dwelling in their outside greenhouse through promoting their produce to nearby markets and restaurants. In the type room, educators are realizing the wonderful applications that hydroponics can have to train kids about science and gardening.

The pace of hydroponic research is increasing at exponential costs as the many advantages are realized. two Associated disciplines such as aeroponics and aquaponics lead the way and no one is aware of what the future holds for such a thrilling inexperienced technology. General Hydroponics will proceed to power innovation and provide reducing side applied sciences and resources.

Part 1 - HYDROPONICS

Chapter 1 - BASIC CONCEPTS

There are many advantages to hydroponic gardening:

Plants grow faster. Experts advocate that plants develop at least 20 percent quicker in hydroponic structures than they do in soil.

Yields are 20 to 25 percent greater with hydroponic systems, in contrast to developing in soil.

No soil is required, which can be a distinct benefit in areas the place present garden soil is poor, or for condominium dwellers were growing in soil is inconvenient.

Hydroponic developing takes less space. Plants don't want to develop sizable root systems to achieve the vitamins they need, so plants can be packed collectively closely—another benefit for those who must garden indoors.

Water is saved. The reservoirs used in hydroponics are enclosed to stop evaporation, and the systems are sealed. This permits vegetation to take up only the water they need.

The first step to placing up your first hydroponic backyard is selecting a gadget that excellent suits your needs from among numerous options. Important elements to reflect on consideration include: how a lot area you have, what you desire to develop and how much, cost, and how an awful lot reachable time you have to spend retaining the system.

The three most simple setups endorsed for novices are the wick, water culture, and ebb and flow. All three of these systems can be built from person factors bought separately, or you can purchase a whole setup kit from online outlets or in a hydroponics store.

Wick Systems

Wick structures are the simplest gadget mechanically, and the best to set up due to the fact there are no transferring parts. The system consists of a reservoir stuffed with water and nutrients, and above it, there is a container crammed with a developing medium. The two containers are linked by using a wick, which draws the nutrient-filled water up into the growing medium, where it is absorbed by way of the roots of your plants. This machine is outstanding for learning the basics, but it may not work nicely with massive vegetation or with water-hungry plants such as lettuce, because the wick cannot provide water quickly enough. However, this machine works extremely nicely with microgreens, herbs, and peppers.

Water Culture

A water subculture gadget is another extremely easy machine to set up. In this system, the flora is placed into a styrofoam platform that sits right on the pinnacle of the reservoir retaining the solution of water and nutrients. A bubbler air pump is delivered to the reservoir to supply oxygen to the plant roots. This system is ideally perfect for water-hungry flora but is not so properly ideal for more long-lived flowers such as tomatoes.

Ebb and Flow

Ebb and waft systems are slightly extra complicated in design, but they are extraordinarily versatile. This machine works through flooding the growing medium with a water/nutrient answer and then draining it lower back into the reservoir. To do this, the gadget requires a submersible pump with a timer. One of the biggest advantages of ebb and flow is that you can use the timer to personalize your plants' watering schedule primarily based on the plant size, variety of plants, temperature, humidity, etc. You also have the alternative of potting plant life in my view for effortless customization or filling the complete tray with growing medium and planting without delay in the tray.

Choosing What to Grow

Just about any plant can be grown hydroponically, however for beginners, it is exceptional to start small. The first-class selections are herbs and veggies that develop quickly, require little maintenance, and do not need a vast vary of nutrients. Fast-growing flora is high-quality seeing that they make it easy to examine how well your system works and tweak it as necessary. It can be an actual letdown to wait months until harvest time solely to discover out your device is not working properly. Maintenance-free plant life is wonderful for novices due to the fact they permit you to focal point on learning about your system—you can cross on to greater complex vegetables later. If you are developing a variety of plants, it is additionally essential to make certain that they are comparable in their nutrient requirements, so that they grow nicely together.

Lighting

Hydroponic structures are frequently indoor structures positioned in locations the place there isn't always get admission to direct sunlight all day long. Most suitable for eating flora require at least six hours of daylight each day, with 12 to sixteen hours even better. So until you have a sunroom or other such area with lots of window exposure, you may in all likelihood want to supply supplemental develop lights. Hydroponic kit systems usually come with the fundamental light fixtures, however, if you are piecing together your very own components, you will need to buy separate lighting fixtures.

The quality lighting for a hydroponics device is HID (High-Intensity Discharge) light fixtures, which can consist of both HPS (High-Pressure Sodium) or MH (Metal Halide) bulbs. The light from HPS bulbs emits a greater orange/red light, which is terrific for flora in the vegetative growth stage.

T5 is every other type of lighting used in hydroponic develop rooms. It produces a high-output fluorescent mild with low warmth and low power consumption. It is best for developing plant cuttings and vegetation with brief boom cycles.

Make certain to put your lighting gadget on a timer so that the lights come on and go off at an equal time every day.

Room Conditions

A hydroponic device must be set up in the proper conditions. Key factors consist of relative humidity, temperature, CO_2 levels, and air circulation. The perfect humidity for a hydroponic grow room is from 40 to 60 percent relative humidity. Higher humidity levels—especially in rooms with bad air circulation—can lead to powdery mildew and other fungal problems.

Ideal temperatures are between 68 and 70 F. High temperatures may cause flora to end up stunted, and if the water temperature receives too high, it might also lead to root rot.

Your grow room needs to also have a sufficient supply of carbon dioxide (CO2). The fine way to make sure this is with the aid of making sure the room has a regular float of air. More superior hydroponic gardeners may complement CO2 levels in the room, because the greater the CO2 available, the quicker your flora will grow.

Water Quality

Two factors can affect water's capability to deliver dissolved nutrients to your plants: the stage of mineral salts in the water, as measured utilizing PPM; and the pH of the water. "Hard" water that carries an excessive mineral content will no longer dissolve vitamins as correctly as water with a lower mineral content, so you may also want to filter your water if it is excessive in mineral content. The perfect pH degree for water used in a hydroponic gadget is between 5.8 and 6.2 (slightly acidic). If your water would not meet this level, chemical compounds can be used to alter the pH into the ideal range.

Nutrients

The nutrients/fertilizers used in hydroponic systems are available in each liquid and dry varieties and both organic and synthetic types. Either type can be dissolved into water to create the nutrient combination required by the hydroponic system. The product you use need to encompass each the essential macronutrients—nitrogen, potassium, phosphorus, calcium, and magnesium—as nicely as the vital micronutrients, which encompass trace amounts of iron, manganese, boron, zinc, copper, molybdenum, and chlorine.

Many nutrient/fertilizers are available that are designed for hydroponic gardening, and you should have properly outcomes if you use them following bundle directions. But keep away from the usage of trendy garden fertilizers in a hydroponic system, as their formulas are designed for use in backyard soil, no longer hydroponic systems.

Choose hydroponic nutrient products that are designed for your unique needs. For example, some are marketed as being great applicable for flowering plants, while others are fantastic for merchandising vegetative growth, such as the greenery of leafy vegetables.

Additional Equipment

In addition to the simple hydroponic setup, it is a suitable concept for beginners to invest in a few additional items.

You will need meters to take a look at the PPM and pH of the water, as nicely as the temperature and relative humidity of the room. There are some aggregate meters available that will test the pH, PPM, and water temperature. You can also buy meters that measure the temperature and/or the humidity in your develop room.

Depending on your climate, you may additionally need a humidifier or dehumidifier to regulate the relative humidity in the grow room to a top-quality level.

You may additionally choose to have some sort of fan or air circulation tool to enhance the air waft in your develop room. Even a simple oscillating fan works well, but as you get greater experienced, you may additionally prefer to invest in a greater state-of-the-art intake-and-exhaust system.

Good Starter Plants

Some flora that work very well for beginners just mastering the fundamentals of hydroponic gardening include:

- Greens such as lettuce, spinach, Swiss chard, and kale
- Herbs such as basil, parsley, oregano, cilantro and mint
- Tomatoes
- Strawberries
- Hot Peppers
- Systems For More Advanced Gardeners

Two greater tricky structures are nice reserved for hydroponic gardeners who have already realized the basics: the N.F.T. system, and the aeroponic system.

N.F.T. stands for the Nutrient Film Technique. It makes use of a regular flow of water/nutrient solution that flows continuously in a loop from a reservoir via a developing tray, where plant roots are suspended in air and soak up nutrients as the solution flows by. But if something goes wrong with the pump mechanism, the roots can dry quickly when the waft stops. This machine requires a consumer who can monitor the equipment and restoration it shortly if troubles arise.

An aeroponic system is a high-tech method in which plant roots are suspended in air and are misted every few minutes with a water/nutrient solution. It is a particularly tremendous method but one that requires sophisticated pumps and misters. If the tools have problems, the plant roots on will dry out and die very quickly.

Chapter 2 - WHY HYDROPONICS

Hydroponics vs Soil. Obviously, there is extra than one difference between the, and in this chapter, we're going to go over the differences between developing flora hydroponically or in soil indoors; how a good deal they yield, how taste and aroma is affected and how everyday plant growth differs.

1. Space Savings

Hydroponics saves a gorgeous amount of space compared to usual soil gardening. Usually, a plant's roots want an area to spread out via the soil. Not anymore! Instead, they are submerged in a bath of oxygenated nutrient solution.

Hydroponics Saves Space.

Vertical Stacking of Lettuce – Soil Can't Do That!

Imagine if you had everything you required to dine in a bit pill.

You didn't want to hunt around for meals or consume three meals a day – you without a doubt popped the tablet and your physique was once dosed with a perfect grant of nutrients.

This is what hydroponics gives your plants. Instead of the usage of soil as a service for the nutrients your flora need, hydroponics makes use of a personalized nutrient answer to surround your plants with flawlessly calibrated vitamin all of the time.

Because of this, you get to pack your vegetation nearer together, ensuing in a huge area savings!

2. Hydroponics Saves Water

Let's suppose about how the common soil gardener waters their plants. Usually every few days they dump an exact amount of water into their soil, making sure correct penetration into the soil so the roots can suck it up.

Sounds great, right?

Well, it's only the phase of the picture.

Some of that water drips out of the bottom of their container or seeps also into the ground. Some of it evaporates out of the soil.

Only a small proportion of the water is used by using the plant. Hydroponics solves this trouble with the aid of using what is referred to as a recirculating nutrient reservoir in most kinds of systems (Deep Water Culture is one of the most popular).

This means that a plant's roots will only take up the amount of water they need at any one time and depart the rest in the reservoir for later. The reservoir is blanketed to prevent evaporation and no water can seep out of the bottom.

This approves the same quantity of water that was used to water a plant in soil for a day to water a plant in a hydroponics set up for days or weeks at a time. You can retailer round 90% of the water used in soil gardening definitely with the aid of switching to a hydroponic setup.

3. No Weeding Necessary

One of the most common excuses I hear when anybody tells me why they don't desire to garden is:

I don't want to spend all of my time on my palms and knees weeding!

Easy solution. Switch to hydroponics. No soil, no weeds. Simple as that.

4. Less Pests and Diseases - Hydroponic Pests

No Soil = No More of These Bad Boys

Following that identical logic, pests and illnesses are appreciably decreased in hydroponics. Soil is taken out of the picture and changed with one of the common hydroponic growing media. Eliminating soil additionally eliminates a lot of the specific soil-borne diseases and pests that plague usual gardening.

5. Double-Headed Time Savings

This is my favorite cause of all. Not only does developing hydroponically save you the time of weeding, pest control, and watering, it also speeds up the growth of the plant.

If you're growing outdoors, that means you get to squeeze in more harvest cycles before your developing season ends.

You also get to observe the growth of plant life at a faster pace and study about all of the unique things you could do to improve the boom a whole lot quicker.

For example, you can take ahead of lettuce from seedling to harvest in around a month in hydroponics compared to two months in soil. Imagine how a good deal faster you could come to be a gardening professional with a time financial savings like that!

6. Gives You EXTREME Control

All of the motives above combine to form one uber-powerful mega motive why hydroponics (and all soilless growing, for that matter) dominates soil gardening: control.

You become the grasp of your plant's environment. It's up to you to create the best nutrient mixture, temperature, humidity, and growing schedule.

It's a form of like that film "The Truman Show." You're the showrunner, and your flora is Truman. You flip the solar on and off. You control when your flowers get fed and what they eat. You're responsible for their well-being. It's a terrific thing!

7. You Get To Become a Guerilla Scientist

Hydroponic Scientist

Run Your Mini-Lab With a Hydroponics System

All of the extra manipulate you have over your developing environment makes for a gorgeous way to research how to develop plants. You can tweak the variables and see however your plants react. You get to personalize the "environmental recipe" to something plant you're growing.

Chapter 3 - WHAT HYDROPONICS IS

What Is The Best Growing Medium For Hydroponics?

When you suppose about plant cultivation, you probable visualize flowers developing in nutrient-rich soil. But with hydroponics, you don't use soil. Rather, the plants are fed via a water-based mineral nutrient solution. However, they still want a growing medium, that is, material to develop in, also acknowledged as the substrate.

The high-quality developing medium for a hydroponics gadget will depend on the kind of machine you choose. The most common preferences of developing media are:

- Grains and Pebbles
- Lightweight expanded clay aggregate
- Perlite
- Vermiculite
- Rockwool
- Oasis cubes
- Coco coir
- Water-absorbing polymer crystals

This chapter will explain all you want to comprehend about selecting a growing medium for your hydroponics system. I will explain why some growing media are nice applicable to particular kinds of hydroponics systems, and explain the elements of all the most common options.

What Role Does The Growing Medium Play In Hydroponics?

In addition to supporting the plant's weight, the medium helps supply moisture and oxygen to the root system and offers the plant with most publicity to the vitamins it needs.

What area unit the benefits To employing a Growing Medium aside from Soil?

One essential advantage is that it eliminates the hazard of pests and diseases frequently located in soil. It also capability you can grow produce in places the place the soil is of bad high-quality – or doesn't exist at all, like on patios, rooftops, and even indoors.

Also, no weeds! And of course, it permits you to have total manipulate of the growing environment, from the temperature to the moisture and oxygen degrees to the nutrients. Best of all, your flora will develop faster and more healthy when their roots don't have to use up power looking out for the nutrients they want in soil.

The Importance of selecting the correct Growing Medium For agriculture

Many types of media can be used for growing plants. For example, you're likely already familiar with peat moss.

Growing media used in hydroponics include inorganic supplies like sand, gravel, and grow stones made of recycled glass, natural materials such as pine bark and coconut fiber, and even air.

When you are deciding on which medium to use for your hydroponic project, you'll need to consider elements ranging from the kinds of flora you intend to develop to the costs and availability of the unique media. However, the main component will be the type of gadget you decide to build, alongside with its design.

Even even though the hydroponic systems and media you can use are all very different, the purpose is usually the same: you want the plant roots to have moisture, but now not too much. If the medium is constantly saturated with water, the roots can suffocate from lack of oxygen, leading to root rot that will kill the plant.

What Are The Main Forms Of Growing Media Used In Hydroponics?

Hydroponics growing media can be categorized into three primary forms: grains and pebbles, foam matrix, and fibrous natural matter. Each form can be used for particular or for conventional functions and may additionally be particularly acceptable for a positive kind of hydroponic growing system.

Let's take a look at the advantages of clay pebbles.

Retains Moisture: Hydroton clay pebbles are super in preserving moisture. When you are attempting to do planting in a water shortage location, clay pebble can make the most out of your irrigation facility. It is an excellent way to hold the water and hold your flowers hydrated along with any intent minerals or vitamins poured in there. It absorbs water and stores its interior for plant life to take in as per their needs. Undoubtedly, clay pebbles are one of the most famous resources when it comes to hydroponics.

Increases Aeration: Sometimes, vegetation suffocate and locate it challenging to develop under the soil. Clay pebbles are light-weight and porous which holds air in them and increases the aeration for the root machine of the plant. The structural formation of these hydro towns is such that it is mild in weight and has adequate house internal to seize the air and let it launch whenever the flora or harvest wishes it. Plants develop higher when they receive acceptable air, water, and sunlight.

Provides Reliable Drainage: Drainage is a real hassle when speakme about harvest or plantation. In areas the place there is a facility or comfort to make wholesome growing, clay pebbles are super for water drainage. As mentioned above hydro towns are outstanding water absorbers, it collects the extra water and stores it for later use. It prevents roots from being broken due to extra water. Generally, clay pebbles are used as a base layer or alongside to help plants get the appropriate amount of water and air via it.

Environment-Friendly: The main ingredient in the manufacturing of clay pebble is clay, which is one hundred percent natural and environment-friendly. Soil and water are appropriate blended to warmness up in an excessive burning furnace and come to be porous tiny balls, which are lighter in weight however have more than a few optimistic properties. There is no involvement of any detrimental gasoline or factor in this procedure for this reason the end product is also entirely nature-friendly. It has no damaging effect and it is full of minerals and natural elements to help plants grow healthy and faster.

Long Life Cycle: One of the most super information about clay pebble is that they closing longer. You can use these clay pebbles for more than one time to plant your preferred bushes or vegetable. These clay pebbles are the best gardening substrate for soilless growing. Unless there is fundamental salt deposition or natural built upon its surface, you can always wash and reuse it. There is no anticipated expiry date for clay pebbles; their lifespan depends on the usage.

Why clay pebbles are one of our pinnacle selections for small growers

Clay pebbles or hydroton (sometimes cited as LECA—light expanded clay aggregate) may be a agriculture substrate with units regarding the dimension of marbles or peanuts. Because they're so lightweight, convenient for transplanting and harvesting, and easy on the hands, they're a preferred of small producers the usage of media bed or Dutch bucket techniques. Clay pebbles are often utilized in each agriculture and aquaponic systems.

Read on for the professionals and cons of the usage of multiplied clay pebbles like hydroton in your hydroponic or aquaponic systems.

Pros of hydroton

1) High pore space means fewer blockages

Larger aggregates like hydroton, pea gravel, and crushed granite have lots of larger area between each rock or pebble than perlite, sand, and other small particles. While the biological floor vicinity isn't typically as high, the pore space is a good deal higher.

What does that mean? Larger pore areas suggest better percolation (flow of answer via the media), even when biofilms from algae and microbes cover the surfaces of the media, and even if some particles are captured in the pore spaces. Hydroton not often become clogged or blocked, so water drains very effectively. This makes it a terrific choice for ebb-and-flow systems and aquaponics media bed systems.

2) Some air-holding capability to hold root zones oxygenated

While it can't rival perlite's air-holding potential (AHC), this grow media does have some capacity to keep air bubbles. Combined with top-notch percolation, hydroton's AHC makes it challenging for frustrating anaerobic zones to occur.

3) Fairly renewable & environment-friendly

Not a whole lot of clay is used to make a cubic foot of hydroton, and clay is abundant, so most human beings reflect on consideration on it an environmentally-friendly medium to use. Compared to many media used in increased quantities that are more traumatic of the earth's supply, hydroton is very pleasant to the environment.

4) Reusable

Although hydroton is a mineral and not regarded as a pollutant, we still don't desire it to quit up in a landfill. Luckily, they are reusable nearly indefinitely. You usually want to rinse any constructed up silt or natural depend on it earlier than reusing it, but except you have a severe salt build up in it, you can reuse it many times.

5) Easy to plant and harvest

Hydroton may be a loose media, so it's easy to transplant and pull plants out of after harvest. Don't underestimate how much time this can save you in wrestling with plant roots and setting apart root balls from the media surrounding them.

Hydroton is a loose media, so it's effortless to transplant and pull vegetation out of after harvest.

Hydroton is a free media, so it's easy to transplant and pull flowers out of after harvest.

6) Good colonization for microbial populations

While grow stones are smoother than some media, they are not so smooth as to discourage colonization using microbes. As you can also understand from our organic floor area resources, BSA affords habitat for the microbes which make vitamins from natural sources like fish feed accessible to plants. Less BSA means fewer microbes, which skill a much less responsive and much less stable system. Though possessing much less BSA than some media, this grow medium nevertheless gives excessive BSA.

Cons of hydroton

1) Water maintaining potential leaves something to be desired

Clay pebbles don't have correct water protecting capacity, or WHC. Since WHC is what approves a substrate to stay moist even after being drained, low WHC capability that crops can get dry and wilted if no longer watered frequently enough. In some structures (with cooler climates, drought-tolerant crops, and/or steady irrigation) this is now not an issue. Growers who have excessive transpiration rates, water-needy crops, etc. will want to parent out a way to maintain the substrate moist.

Low WHC isn't a massive deal for most producers; just be aware of it and make positive you have time-honored adequate watering.

2) Fairly costly

Hydroton is extremely easy to work with, which makes it the first preference for many small growers, however, it's a bit too pricey for most giant growers to use it.

3) Can reason troubles with pumps and plumbing

Because hydroton floats for the first few months until it's been saturated, the pebbles can get sucked into filters or drain traces and motive blockages.

Pros of perlite

What is perlite?

Perlite is ore that has been superheated in a kiln till it expands like popcorn. This makes it terribly light-weight and offers it air maintaining capacity—a real advantage for growers making an attempt to keep up root zones aerated.

Perlite has been used for many years in insulation, cement, and building materials, however lately has been used more and more for matters like filtering and as a growing substrate. Many hydroponic growers use perlite as their important medium.

Pros of perlite
1) Perlite is usually reusable.

The only time you would possibly throw perlite away is if you have a very bad disorder problem and no desirable way to sterilize it.

If you have a horrific infection of something like pythium, then you'll want to wipe out any inoculum (the infecting phase of a disease, for example, the spores) before the use of that media with new crops. You can sterilize the media with heat (using a rented soil sterilizer) or with chemicals with a oxide answer or bleach answer (remember to rinse it extraordinarily nicely if the usage of bleach).

2) Perlite helps deal with anaerobic conditions.

Because perlite holds air so properly and because it's a coarse texture, it can do wonders for systems dealing with oxygen issues. A lack of oxygen in water, soil, or anywhere that roots are growing (the root zone) causes anaerobic conditions. This permits anaerobic microorganism (decomposers) to come in and start doing their thing. Since their factor is decomposing, this is very horrific news for plant roots. The point is: averting anaerobic zones is crucial!

Growers can avoid anaerobic zones with the aid of maintaining the water oxygenated (use correct glide rates, turbulence, and air stones) and fending off build-up and compaction in the growing medium. Perlite is a big assist in this area, as the giant particle measurement now not solely presents air pockets and has no compaction problems, but without a doubt has some oxygen-holding potential and exchange.

3) Perlite is inexpensive.

You can get four cubic toes of perlite for $14 bucks at a hydroponic or greenhouse shop (or online), while different sterile pH-neutral media, like hydroton, can value almost twice that. Since most applications (in soilless mixes or structures like Bato buckets) use an exceptionally small quantity of the medium, perlite is fairly cost-effective.

4) Sterile and pH neutral

Some soilless media ought to be sterilized before being used to keep away from the introduction of pests and ailments into the system. Since perlite is now not sourced from a natural source (similar to coir and peat) and has been sterilized in the introduction process (being superheated), it has had nearly no chance for bacterial, fungal, or insect pests to get into it. This helps you avoid pest problems!

Perlite is additionally pH neutral, unlike some media like expanded shale and rock wool. Those sorts of media can be mildly basic, which influences machine pH and intricate pH dosing. A neutral media can be tons more convenient and higher for long-term device health.

Cons of perlite

1) Perlite ore is not a renewable resource.

There is solely so much ore in the world. Although it'll eventually renew, it's not renewable in human time. That said, we don't use that a good deal of it relative to all that is there (in 60 years we've used much less than 1% of the world's perlite ore) and it's inexpensive. You'll have to weigh the professionals and cons of that to decide whether you suppose this is sustainable or not.

2) Aggressive root structures can motive blockage.

With massive aggregates like hydroton, plant roots developing down into the pore area isn't going to affect the percolation much. After all, there's pore space to spare between the fairly massive particles.

Perlite, on the different hand, is composed of smaller particles. This ability that when vegetation with an aggressive root machine (either very mature plants or flowers like mint and chives with a lot of roots) prolong into perlite, the pore spaces can get mucked up and blocked.

pros and cons of perlite in hydroponics

Roots can fill up small pore areas (like those in vermiculite), inflicting clogging, debris build-up, and pooling water.

3) Vulnerable to solids loading

In addition to being filled with plant roots, the air pores in perlite can capture solids like algae, debris, and biofilm, with a similar result: blocked percolation.

This isn't normally a huge problem except the device is pretty dirty or is being run on an organic hydroponic solution. This type of answer relies on a sturdy neighborhood of microbes to cycle nutrients, and so a lot of thicker biofilm will shape on the floor of the substrate.

4) Mud hurts fish and may be dangerous if inhaled

Don't use perlite with fish! If you appear at perlite under a microscope, it looks like a collection of small glass bubbles. And in fact, that's what it is. That capacity that even though picking up perlite with your fingers won't cut you, it is abrasive and can cause real damage to gentle touchy tissue—like the gills of a fish, and your throat and lungs!

That being said, don't use perlite in an aquaponic operation. Wear a mask or respirator when you're coping with dry perlite to keep away from respiratory in perlite dust. Once it's wetted down, the dust shouldn't be a problem.

Some of the advantages of vermiculite include:

Vermiculite is lighter than soil mediums. This is especially necessary when growing indoors and when developing at peak such as roof gardens, tiered greenhouse house & grow rooms. A lighter medium additionally has the advantage of being simpler to carry, store and manage in general

Vermiculite offers a sustainable answer to growing plants. While peat has been used in the past, peat is no longer a practical way of developing plant life due to the negative outcomes of peat harvesting from our bogs. Bogs provide a special and natural habitat in which a total host of wild birds, insects, and animals live. Continuous depletion of peat from bathrooms is putting all these animals in danger.

Vermiculite is, in reality, useful to flowers and their roots. By adding vermiculite to the soil you can improve the soil's capability to keep and release water back to the plants. Vermiculite is also porous and so will preserve excessive degrees of air in soils which is imperative for both root breathing and the existence of soil microorganisms.

Vermiculite can be used in hydroponics and soilless growing systems. These are beneficial in that little or no water or vitamins are wasted or washed away. Hydroponic growing systems work in a wide variety of ways. Essentially they contain a machine of sitting plant's roots in a pot that is flooded with water and then enables to drain free. The advantages right here consist of the reuse of water and nutrients as nicely as full manage over watering & nutrient tiers that the flowers receive

The endless different makes use of vermiculite imply that every gardener needs to have a bag at their disposal at any time. Vermiculite will be utilized in storing bulbs and vegetables- keeping them dry and funky. It can also be used when planting spring and summer season.

How to Use Rockwool as Medium and also the edges It Offers

Many people use Rockwool as a developing medium for hydroponic gardening. That's due to the fact there are several advantages for growers if they research to use it correctly. Hydroponic systems help a large number of growers of flowers and plant life all over the world. For one, they produce consequences for several instances faster than soil. It also lets in humans to plant and grow all yr around.

That is very essential for those that stay in locations where there are cold temperatures most of the year. Because in hydroponics, anything developing medium you are the usage of replaces soil or dirt, what you use is just as important. A lot of hydroponic customers are now turning to Rockwool for developing flowers and flowers. That's because Rockwool presents a lot of benefits and away better results. Still, there are a few matters to think about earlier than you use this method.

Advantages Of Rockwool

One of the many benefits hydroponic growers take away from the usage of Rockwool for developing is yields. Those that use Rockwool tend to yield crops at a plenty higher and quicker degree than different methods. Perhaps that may also explain why this medium is one of the most frequent strategies used in hydroponics today. Rockwool is in particular composed of limestone and/or granite. Once it is heated and melted, it is then woven into threads.

The melting basaltic rock is spun simply like cotton candy. After it has been spun, the cloth is shaped into different sizes or shapes. Those consist of cubes, flocking, blocks and slabs. Rockwool is non-degradable, porous and sterile. One of the reasons it works so nicely is because it sucks up water rapidly and easily. However, due to the fact of that, you have to be careful about letting it turn out to be saturated. If not, you can quit up suffocating or killing the root of your plants or flowers. Also, it can lead to root rotting or stem rot.

For small growers, Rockwool is very really useful for the reason that it presents a broad variety of situations and systems. Another outstanding benefit is that it is very handy to use and set up as well.

How To Use Rockwool

Although there are numerous benefits to the usage of Rockwool, it does require interest and care. There are a few matters you need to think about before you commence the use of Rockwool.

First, before the use of it, make certain the Rockwool has been pH balanced. The great way to do this is by using without a doubt soaking it in a pH stability water solution before the use of it. Whether you pick massive cubes, slabs or pots of granulated Rockwool, the training is very important. You have to sit down the Rockwool down on an even, flat surface. The drainage holes have to also be set up successfully seeing that now not doing so can have terrible results. Make certain your containers or pots have plenty of drainage holes as well.

The subsequent step is the irrigation software or technique you set up. The number of holes vastly depends on how many plants you have in a slab. Since a Rockwool slab can maintain about four plants, you will need about four drippers. Doing this will make certain that the entire slab will get sufficient irrigation even if one of the drippers will become clogged.

Keep in idea that one of the biggest cons that hydroponic developing systems and the use of Rockwool presents is maintenance. While this method yields results a lot quicker than soil, it requires greater servicing. Identically, mineral wool or stone wool (as Rockwool is also known) wants preservation and attention.

There are a lot of blessings to using hydroponics and developing in Rockwool slabs or cubes. No depend on which technique you choose, possibilities are that you will run into problems at first. However, like most matters in life, experience, trial and error is what counts. You want to screen your setup and see which method yields the first-rate consequences for you.

Oasis Cubes are manufactured from water-absorbent foam, Phenolic foam additionally acknowledged as Floral Foam. Also acknowledged as Oasis Root Cubes, they provide an excellent beginning surrounding for seedlings and plant cuttings, no longer as a full growing medium. Lightweight pre-formed cubes designed for plant propagation. They have a neutral hydrogen ion concentration and retain water well.

The foam is designed with what equates to small capillaries that not only permits for the transpirational pull of moisture however oxygenation as well.

They offer desirable beginning surroundings for seedlings and plant cuttings, and it's the place it ends.

Oasis Cubes have no buffering capacity, no cation trade capacity, and no initial nutrient charge. Beyond seed staring and propagation they are of restricted value. Attempts to use them in raft systems have no longer grew to become out well.

Oasis cubes square measure designed notably for seed propagation in business agriculture producing structures and most typically used for quick germination of crops like lettuce and Cole crops, onions and alliums, herbs and now and then tomato and eggplant seeding. The floral enterprise many times uses it for a wide variety of flowers each annuals and perennials.

Disadvantages:

- Not eco-friendly, they are very similar to Styrofoam in this manner.

- Although they are reusable, the initial fee is pretty high.
- Advantages
- Excellent to begin seeds or propagate cuttings
- pH neutral
- no pre-soaking needed
- Water retention of 30 - 40 times their very own weight.
- Accelerates germination
- Enhances root improvement in early degrees of plant life existence cycle

What Is Coco Coir And How Is It Made?

Coco coir is a byproduct of coconut fiber. It was once first used in gardening in the West in the 19th century but fell out of desire due to the fact the low-quality coco available at the time degraded when used for temporary growing. Toward the cease of the 20th century, it was once rediscovered as an organic, environmentally sustainable substrate when new production strategies made it possible to create hardier products.

Coco coir is manufactured the usage of fiber that's torn from coconut shells. The tiny grains of coir are extracted from the coconut shell and pulverized into a packable developing substrate. First, the coconuts go through the retting process, a curing approach that naturally decomposes the husk's pulp. Traditionally, coconut husks had been immersed in water for six months or longer to decompose. Today, the retting method can be done in a little over a week with the use of modern mechanical techniques.

Next, the coconut fiber is removed from the shells through steel combs, in a procedure recognized as defibering.

Once the fiber, or coir, is gathered from the husk, it's then dried, pressed into bricks, discs, coir pots. or bag as free mulch. In this dried, processed state, the coir is geared up to promote and use.

Basic Types Of Coco Coir

There are three primary types of processed coco coir: pith, fiber or chips. Using a combination of the exclusive types has its benefits.

Coco pith, or peat, appearance almost like bog moss however could be a made, brown color. The density of this product capability it retains water extremely properly — so for this reason, you would possibly not want to use simply coco peat, due to the fact it should swamp the roots of your plants.

Coco fibers are stringy bundles that enable oxygen to without difficulty penetrate a plant's root system. By itself, the fiber is now not very absorbent and will wreck down over time, which decreases how a great deal air receives to the roots of your plants. However, it is hardy adequate for reuse.

Coco chips are small chunks of coir that combine the fine residences of the peat and fiber. Coco chips preserve water well, however additionally enable for air pockets, too.

If you're an experienced grower, you can put together your combinations from these distinctive sorts of coco coir, however, agencies furnish premixed products to eliminate all the problems of doing it yourself. Dried bricks are common — all you have to do is add water — however, most coco in brick structure tends to be of a decrease unprocessed quality.

The Benefits Of Using Coconut Coir

Let's take a second to cover the professionals of this growing medium.

Quick harvests and huge yields: When used for drain-to-waste growing, coco coir gives top-notch results. With the proper coco coir vitamins in your water bath, your plants spend much less time searching for meals and more time growing. Learn greater about the usage of the right coco coir nutrients here.

Plenty of room for the root system: Coco coir affords an uncommon combination of awesome water retention, reliable drainage, and ideal aeration. It gives the roots masses of room, permitting for most reliable air exposure.

pH-neutral value: Coco coir has an impartial pH range of 5.2–6.8, but you'll still want nutrients to assist because this vary will fluctuate over time. Learn why maintaining a balanced coco coir pH is so necessary here.

Minimizes damaging pathogens and reduces the hazard of pests: This medium boasts antifungal properties, which maintains the roots happy. It can repel some pests, which means your develop is easier to maintain. (If you've experienced plant pests or illnesses in the past, here are some plant protection guidelines to assist up your game.)

Environmentally aware product: On average, a coconut tree produces one hundred fifty coconuts annually. Coco coir makes use of parts of the fruit that used to go to waste.

Reusable medium: When accurate treated, coco coir can be reused. It's durable, but you want to make positive you prep it efficiently for the next growth cycle to warranty a hearty crop.

What Are The Drawbacks Of Using Coco Coir?

Any develop medium has its limitations, and you have to recognize the characteristics of coco coir to ensure you strengthen the first-class crop possible.

Possible excessive salt content: Make sure you research how the coco medium you pick out is produced. If the husks have been soaked in saltwater, confirm it used to be rinsed with fresh water utilizing the manufacturer, or learn how to appropriate do it yourself.

Chemical treatment: At the top of the drying process, coir bales might be treated with chemical agents to make certain pathogens didn't bloom inside. Learning how it was treated can also help you manage your crop because the chemical residue could affect plant growth. Read the product label or refer to the manufacturer's internet site to examine more.

It can lock out calcium, magnesium, and iron: Because of its high cation change rate, coco coir shops, and releases nutrients as needed, but it tends to maintain calcium, magnesium, and iron. This means you'll need to use unique coco coir nutrients to raise Ca, Mg and Fe range for healthy crops.

Coco Coir Features That May Be A Pro Or A Con

Coco needs to be fed daily. To overcome the cation exchangeability of the coco, it is nevertheless essential to use a coco-specific nutrient, but you additionally want to feed faster than the coco can negatively react with the nutrients. Coco is extraordinarily challenging to overwater, conserving on to oxygen even when drenched, so some hand-watering soil growers might also discover coco requires more work. However, industrial growers often love this feature due to the fact they can connect automated drip lines to the plants.

Use Advanced Nutrients For coco palm fiber Grow to urge the simplest Results.

Because of the complexities of the coco coir medium, you have to use dependable nutrients to protect your crops. Thankfully, the 25 Ph.D.'s at our lab have found the missing link to release coco coir's growing potential.

Most vitamins on the market deliver extra Ca and Mg for coco coir growing. But our researchers have determined that the lacking piece of the coco puzzle is iron. Not only do your flowers want more Ca and Mg when the usage of coir, but they additionally want extra iron because the coir also chemically binds to iron. If you've used popular coco fertilizers in the past, your flowers possibly struggled and produced a disappointing yield.

What Are Hydroponic Hydrogels?

Hydrogels, additionally called hydrophilic gels, have been used seeing that the Seventies in horticulture. Before the '70s, they had been made of herbal materials earlier than being synthetically engineered with three-dimensional, ultra-absorbent polymers, generally proteins such as gelatin and collagen, and polysaccharides like agarose, alginate, and starch. Water can be absorbed at various hundred instances the structure's weight thanks to the strong polymeric backbone inside the hydrogel.

Once water is absorbed, it can seep with regulation into the surrounding environment, making hydrogels a beautiful tool for gardeners. Not solely can they soak up water, they can additionally absorb liquid nutrients that are then launched predictably.

What Makes Hydroponic Hydrogels so Useful?

So, the science is cool, however, how can it be virtually utilized for gardening purposes? For starters, in areas the place water is challenging to come with the aid of or there is drought, a slow launch of water will minimize evaporation, allowing more of the water to advantage flowers while conserving resources.

Also, even the most enthusiastic gardeners enjoy some vacation time. Employing hydrogels creates a worry-free approach of preserving plant life hydrated while you're away besides having to hassle the neighbors.

When discussing the future of growing, hydrogel water is being used on the International Space Station (ISS). Hydroponics is brilliant for developing sparkling produce for astronauts, however, water can be cumbersome in the tight quarters of the ISS. This is certain to be studied greater as plans are being fleshed out to put humans on Mars and other house explorations.

<u>Benefits of Hydroponic Hydrogels</u>

- When it comes to your develop set-up, there are a few benefits that can be won from the usage of hydrogel water or gel crystals.
- Water is slowly released
- Nutrients can be slowly released
- Easily replenished
- Conserves water
- Good for people that forget to water or can't water plant life regularly
- Can be introduced to soil or used in a hydroponic system
- Can be an enjoyable way to get children concerned with developing produce
- Future applications are exciting
- Drawbacks of Hydroponic Hydrogels
- When it comes to drawbacks, there virtually is only one: they launch a limited amount of moisture. Hydrogel water and gel crystals work as an alternative properly with seedlings, grass, leafy greens, and other comparable plants due to the fact they can maintain up with the essential moisture grant wished for these types of plants.

- However, hydrogels simply don't have the capacity, at least not yet, to water closely fruited vegetation like tomatoes and peppers, which require much more water to grow. Large plants would additionally pose a challenge.
- This may alternate in the future, however, as scientists are already working on ways to improve the usage of these materials for growing. In Japan, high-tech polymers were used in a thin sheet to reduce the problems with growth in this fashion and maximize the benefits.
- This test labored as a way to spread nutrients and encourage root growth while working as a medium. Scientists grew tomatoes, melons, and spinach the use of this method, but only the smaller flora had been successful. While no longer a whole win, it's a step in the right direction.

Chapter 4 - HYDROPONICS SYSTEMS

It can be very puzzling to get started in hydroponics. Figuring out however it all works, the way to opt for a system, what to grow, and even the way to grow square measure all difficult.

Types of Hydroponic Systems

There are six essential sorts of hydroponic structures to choose from:

1. Wick Systems
2. Deep Water Culture (DWC)
3. Nutrient Film Technique (NFT).
4. Ebb and Flow (Flood and Drain)
5. Aeroponics
6. Drip Systems
7. Wicking Systems

A wicking device is the most basic kind of hydro gadget you can build. It's been used for lots of years, even though it wasn't viewed as a hydroponic device back then.

It's what's known as passive hydroponics, which means that you don't want any air pumps or water pumps to use it.

Nutrients and water are moved into a plant's root zone through a wick, which is often something as simple as a rope or piece of felt.

One key to success with a wicking machine is to use a developing media that transports water and vitamins well. Good alternatives consist of coconut coir, perlite, or vermiculite.

Wick systems are precise for smaller plant life that don't use up a lot of water or nutrients. Larger vegetation can also have a difficult time getting ample of both through a simple wick system.

Benefits of Wick Systems

- Truly "hands-off" if you set it up efficaciously
- Fantastic for small plants, beginner gardeners, and children
- Downsides of Wick Systems
- Not proper for larger plants
- Incorrect wick placement or material can suggest dying for your plants
- Deep Water Culture (DWC) Systems

- Deepwater culture, which I will refer to as DWC from here on out, is hands-down the easiest type of hydro device to use.
- In a DWC system, you use a reservoir to keep a nutrient solution. The roots of your plant life are suspended in that solution so they get a constant furnish of water, oxygen, and nutrients.
- To oxygenize the water, you use an air pump with an air stone to pump bubbles into the nutrient solution. This prevents your roots from drowning in the water — a bizarre aspect to suppose about, but it can (and does) happen to many amateur hydroponic gardeners.
- Your flowers are generally housed in net pots that are placed in a foam board or into the top of the container that you're the use of for your reservoir. With some hydroponic growing media introduced into your net pots, they supply a home for the very commencing of your root system and plant stems.

Benefits of Deep Water Culture

- Very cheaper and handy to make at domestic
- Extremely low-maintenance
- Recirculating, so much less wasted inputs
- Downsides of Deep Water Culture
- Does no longer work nicely for giant vegetation
- Does now not work nicely for flora with lengthy developing length

Nutrient Film Technique (NFT) Systems

The Nutrient Film Technique, which I will refer to as NFT, is a famous business hydroponic system.

Plants are grown in channels that have a nutrient solution pumping through them and constantly running alongside the backside of the channel. When the solution reaches the cease of the channel, it drops lower back into the main reservoir and is sent again to the starting of the device. This makes it a recirculating system, simply like deep water culture.

Unlike deep water culture, your flora roots are no longer submerged in an NFT device — consequently the "film" section of the system's name.

Plants are positioned in these channels using internet pots and developing medium and can be replaced or harvested on a one-by-one basis.

<u>Benefits of Nutrient Film Technique</u>

- Minimal growing medium wished
- Recirculating gadget means less waste
- Downsides of Nutrient Film Technique
- Pump failure of any sort can break your crop
- Roots can become overgrown and clog the channels
- Ebb and Flow / Flood and Drain Systems
- Ebb and Flow systems, which are also known with the aid of the identify Flood and Drain, are a less-commonly viewed system. But they're nevertheless quite high quality and can be an excellent preference relying on your situation.

- Unlike the preceding two hydro systems we have covered, an ebb and waft system does no longer expose the roots of your vegetation to nutrient answer consistently.
- Instead, you develop in a tray crammed with a growing medium. The tray is "flooded" with your nutrient answer a few times per day, relying on factors like:
    - The dimension of your vegetation
    - The water requirement of your plant life
    - The air temperature
    - Where your plant life are in their boom cycle
    - And many greater
- Flooding is finished utilizing the use of a reservoir under the tray, a water pump, and a time to schedule the flooding cycle.
- After the tray is flooded, gravity drains the solution back down into the reservoir, the place it is being oxygenated via an air pump and air stone. It sits there waiting for the subsequent flood cycle, and the process goes on.

- Hydroponic growers choose ebb and waft structures for their flexibility. Most of them will fill the tray with a developing medium of their desire and additionally add net pots to arrange their flora and control the roots a bit more.

<u>Benefits of Ebb and Flow</u>

- Efficient use of water and power
- Highly customizable to your particular wishes
- Downsides of Ebb and Flow
- Roots can dry out shortly if environmental conditions are off or the pump or timer fails
- Uses a lot of developing medium
- Aeroponics Systems
- Aeroponic structures are the most "high-tech" hydroponic setups that you can build. But they're no longer that complex as soon as you recognize how they work.
- An aeroponic system is comparable to an NFT device in that the roots are more often than not suspended in the air. The difference is that an aeroponic device achieves this with the aid of

misting the root region with a nutrient answer constantly as a substitute for walking a thin film of nutrient solution along a channel.
- Some growers select to mist on a cycle like an ebb and waft system, however, the cycle is a whole lot shorter, commonly solely waiting a few minutes between each misting. It's additionally viable to mist on a continual groundwork and use a finer sprayer to make certain more oxygen receives to the root zone.
- Aeroponic structures have been proven to grow plants even quicker than some of the less complicated systems like deep water culture, however, this has not been validated to be real in all cases. If you choose to scan with this system, you will want specialized spray nozzles to atomize the nutrient solution.

Benefits of Aeroponics

- Roots regularly are uncovered to extra oxygen than submerged-root systems
- Downsides of Aeroponics
- High-pressure nozzles will fail and roots will dry out

- Not as low priced or convenient to set up as different techniques
- Drip Systems
- Drip systems are extremely frequent in business operations, however much less frequent in leisure gardens. This is because they're simple to operate a giant scale but slightly overkill for a smaller garden. Regardless, they're an exquisite way to grow hydroponically that you should consider.

Benefits of Drip Systems

- High degree of control over feeding and watering schedule
- Less likely to destroy
- Relatively affordable
- Downsides of Drip Systems
- May be overkill for a smaller backyard
- Fluctuating pH and nutrient levels (if using recirculating system)
- High waste (if the usage of waste system)
- Well, there you have it. The six essential types of hydroponic systems, how they work, and the ups and downs of everyone.

- No rely upon which one you choose, your flowers will develop quick and huge furnished you care for them properly. Hydroponics affords first-rate flexibility, so even if you're experiencing some troubles, you must have no hassle correcting them and getting your plant life lower back on track.

Chapter 5 - HOW TO CHOOSE THE RIGHT HYDROPONIC SYSTEM

Deciding which hydroponic machine you will use will rely upon how a great deal of money you will spend, what variety of flora you are capable to grow, and how profitable your garden will be. Therefore, it is quintessential that you choose a system that fits your budget, needs, and experience. Hydroponic structures vary in what form of equipment is required, how the vitamins are delivered, and what media can be effectively used.

1. Available space

You want to, first of all, check your growing web page to decide the accessible space. This is because the space on hand determines the number of pots or buckets that can be blanketed in a given hydroponics system. This will finally decide the number of plants that you can grow in your developing site.

In most cases, the smaller hydroponics systems require about sixteen rectangular ft of flooring space. You ought to additionally put together a greater area that will be used to keep the water reservoir, lighting, pump, and coolers. Therefore, analyzing your developing site on hand space is a key necessity.

2. Automation

Hydroponics systems have extra factors such as pumps, grow lights, and coolers which are integral in ensuring the best stages are attained when developing plants. In the market, there are both automated and guide structures and it's the responsibility of the gardener to choose the high-quality system that guarantees efficiency.

Recent research has indicated that most indoor gardening failure occurs as a result of bad temperature manage and water levels.

Purchasing an automatic system will give you an easy time when developing plant life because the system will automatically manipulate the required most desirable levels.

The modern structures have electronic units that automatically reveal humidity, temperature, lighting, and water ranges thus, relieving you a lot of manual adjustment burden.

3. Expandability of the Hydroponics System

As a beginner, you would possibly want to try gardening with a small hydroponic device kit and later increase it to develop more plants. Once you are satisfied with the advantages of training this handy and enjoyable gardening method, you can format to extend the system to keep greater plants.

In this case, you need to have enough space that can hold extra buckets or pots to correctly accommodate the extra plants. Expandability of the gadget determines your total output and it is an imperative thing that can help you make the proper choice.

4. Energy efficiency

Every hydroponics machine is operated the use of electrical energy that helps pumping, lighting, and air conditioning. Electricity fees can run high especially when a farmer declines to use energy-saving LED bulbs. A full spectrum of mild is wished during the device to ensure top-quality growth of your plants.

Therefore, when buying your hydroponics system, usually make certain that you use energy-saving LED bulbs and this will go a lengthy way in minimizing your working expenses.

5. System charge and setup costs

Hydroponics structures can be sold as pre-built or the gardener can determine to assemble one. Constructing your personal DIY Hydroponics device at home will require professional carrier the place you will have to rent a professional to set the system in place if you can't do so.

This would possibly be expensive to novices and it requires shut supervision meaning you will have to be there in the course of installation.

On the other hand, the market has a large array of pre-built hydroponics systems that are customized to swimsuit your preference. With the whole lot already set in place, you will solely be required to set the system in your desired vicinity and kick-start your indoor gardening assignment proper away.

Depending on your price range diagram you can be capable to make the right choice whether to DIY or purchase a pre-built system.

Chapter 6 - ADVANTAGES AND DISADVANTAGES OF HYDROPONICS

1. No soils needed

In a sense, you can grow vegetation in places the place the land is limited, doesn't exist, or is closely contaminated. In the 1940s, Hydroponics was successfully used to grant fresh veggies for troops in Wake Island, a refueling give up for Pan American airlines. This is a far-off arable vicinity in the Pacific Ocean. Also, Hydroponics has been viewed as the farming of the future to grow ingredients for astronauts in the area (where there is no soil) via NASA.

2. Make better use of house and location

Because all that plant life need is supplied and maintained in a system, you can grow in your small apartment, or the spare bedrooms as lengthy as you have some spaces.

Plants' roots normally amplify and unfold out in search of foods, and oxygen in the soil. This is now not the case in hydroponics, where the roots are sunk in a tank full of oxygenated nutrient answer and at once contact with necessary minerals. This skill you can grow your vegetation a whole lot closer, and hence big house savings.

3. Climate control

Like in greenhouses, hydroponic growers can have complete control over the climate - temperature, humidity, light intensification, the composition of the air. In this sense, you can grow foods all year spherical regardless of the season. Farmers can produce foods at a fabulous time to maximize their enterprise profits.

4. Hydroponics is water-saving

Plants grown hydroponically can use solely 10% of water in contrast to field-grown ones. In this method, water is recirculated. Plants will take up the necessary water, whilst run-off ones will be captured and return to the system.

Water loss only takes place in two types - evaporation and leaks from the gadget (but an environment-friendly hydroponic setup will reduce or don't have any leaks).

It is estimated that agriculture makes use of up to 80% water on the floor and surface water in the US.

While water will grow to be a fundamental issue in the future when food production is expected to make bigger by 70% according to the FAQ, Hydroponics is considered a workable solution to large-scale meal production.

5. Effective use of nutrients

In Hydroponics, you have a one hundred percent manage of the nutrients (foods) that plants need. Before planting, growers can check what flowers require and the unique amounts of nutrients wished at particular stages and combine them with water accordingly. Nutrients are conserved in the tank, so there are no losses or modifications of vitamins like they are in the soil.

6. pH manage of the solution

All of the minerals are contained in the water. That capacity you can measure and adjust the pH degrees of your water mixture a great deal more effortlessly in contrast to the soils. That ensures the most appropriate nutrients uptake for plants.

7. Better increase rate

Is hydroponically flowers grown quicker than in soil? Yes, it is.

You are your boss that commands the whole environment for your plants' growth - temperature, lights, moisture, and in particular nutrients. Plants are positioned in the best conditions, whilst nutrients are provided at the adequate amounts and come into direct contact with the root systems. Thereby, flowers no longer waste valuable energy looking out for diluted vitamins in the soil. Instead, they shift all of their center of attention on growing and producing fruits.

8. No weeds

If you have grown in the soil, you will understand how anxious weeds cause to your garden. It's one of the most time-consuming duties for gardeners - till, plow, hoe, and so on. Weeds are broadly speaking associated with the soil. So do away with soils, and all bothers of weeds are gone.

9. Fewer pests & diseases

And like weeds, obtaining rids of soils helps create your flora abundant less at risk of soil-borne pests like birds, gophers, groundhogs; and ailments like Fusarium, Pythium, and Rhizoctonia species. Also when growing indoors in a closed system, the gardeners can effortlessly take controls of most surrounding variables.

10. Less use of insecticide, and herbicides

Since you are the usage of no soils and while the weeds, pests, and plant diseases are heavily reduced, there are fewer chemical substances used. This helps you develop cleaner and more healthy foods. The cut of insecticide and herbicides is a sturdy factor of Hydroponics when the criteria for present-day existence and food protection are greater and more positioned on top.

11. Labor and time savers

Besides spending fewer works on tilling, watering, cultivating, and fumigating weeds and pests, you experience an awful lot of time saved due to the fact plants' boom is established to be greater in Hydroponics. When agriculture is deliberate to be extra technology-based, Hydroponics has a room in it.

12. Hydroponics is a stress-relieving hobby

This pastime will put you lower back in contact with nature. Tired after a long working day and commute, you return to your small condominium corner, it is time to lay back the whole thing and play with your hydroponic garden. Reasons like lack of spaces are no longer right. You can start fresh, tasty vegetables, or indispensable herbs in your small closets, and enjoy the relaxing time with your little inexperienced spaces.

Seem like there are lots of benefits of Hydroponics and the picture seems to strive to persuade you into Hydroponic growing. But preserve studying to examine about its downsides.

13. An aquicultural garden needs it slow and commitment

Just like any things worthwhile in life, hard-working and responsible mindset gives first-rate yields. However, In soil-borne counterparts, plant life can be left on its own for days and weeks, and they nevertheless continue to exist in a short time. Mother nature and soils will help regulate if something is not balancing. That's not the case in Hydroponics. Plants will die out more rapidly without ideal care and adequate knowledge. Remember that your plants area unit reckoning on you for his or her survival. You need to take the top care of your plants, and the system upon initial installation. Then you can automate the complete issue later, however, you nonetheless want to gauge and prevent the surprising issues of the operations, and do common maintenance.

14. Experiences and technical knowledge

You are walking a gadget of many sorts of equipment, which requires fundamental particular know-how for the units used, what flowers you can grow and how they can survive and thrive in a soilless environment. Mistakes in placing up the structures and plants' increased ability in this soilless surroundings and you quit up ruining your entire progress.

15. Organic debates

There have been some heated arguments about whether Hydroponics have to be licensed as natural or not. People are questioning whether flowers grown hydroponically will get microbiomes as they are in the soil. But people around the world have grown hydroponic vegetation - lettuces, tomatoes, strawberries, etc. for tens of years, mainly in Australia, Tokyo, Netherland, and the United States. They have provided meals for hundreds of thousands of people. You cannot anticipate perfection from something in life. Even for soil growing, there are nevertheless more dangers of pesticides, pests, etc. compared to Hydroponics. There are some organic growing techniques counseled for Hydroponic growers.

For example, some growers furnish microbiomes for plant life using the use of organic developing media such as coco coir and add worm casting into it. Natural-made nutrients are commonly used such as fishes, bones, alfalfas, cottonseeds, neems, etc.

For this debate for the natural product issue, there will still be researches carried out presently and in the close to future. And we'll know the answer then.

16. Water and electrical energy risks

In a Hydroponic system, more often than not you use water and electricity. Beware of electrical energy in an aggregate of water in shut proximity. Always put protection first when working with the water structures and electric-powered equipment, specifically in industrial greenhouses.

17. System failure threats

You are using electricity to manipulate the complete system. So believe you do not take preliminary movements for a strength outage, the device will stop working immediately, and plants may dry out shortly and will die in countless hours. Hence, a backup electricity supply and sketch ought to continually be planned, mainly for super scale systems.

18. Initial expenses

You are certain to spend under one hundred to a few lots of greenbacks (depending on your backyard scale) to purchase tools for your first installation. Whatever systems you build, you will need containers, lights, a pump, a timer, developing media, nutrients). Once the gadget has been in place, the cost will be decreased to only vitamins and electricity (to hold the water gadget running, and lighting).

19. Long return per investment

If you comply with news on agriculture start-up, you may additionally have recognized that there have been some new indoor hydroponic commercial enterprise started recently. That's an exact element for the agriculture area and the development of Hydroponics as well. However, business growers nevertheless face some massive challenges when starting with Hydroponics on a massive scale. This is mostly due to the fact of the high preliminary prices and the long, uncertain ROI (return on investment). It's not effortless to detail a clear worthwhile sketch to urge for funding while there are additionally many other alluring high-tech fields out there that appear fairly promising for funding.

20. Diseases and pests may additionally unfold quickly

You are growing plant life in a closed system using water. In the case of plant infections or pests, they can amplify quickly to plant life on the equal nutrient reservoir. In most cases, ailments and pests are now not so much of trouble in a small gadget of home growers.

So don't care a great deal about these problems if you are beginners.

It's only intricate for huge hydroponic greenhouses. So better to have an accurate disease administration design beforehand. For example, use just easy disease-free water sources and developing materials; checking the systems periodically, etc.

Should the illnesses happen, you need to sterilize the contaminated water, nutrient, and the whole device fast.

Chapter 7 - SUITABLE CROPS IN HYDROPONICS

TOP 5 PLANTS FOR NEW HYDROPONIC GARDENS

The 5 quality plants to grow in a hydroponic device are:

- Lettuce
- Spinach
- Strawberries
- Bell Peppers
- Herbs

Growers have found that these flora take to hydroponics like a duck to water. They're durable, speedy growing and don't take a lot of work to get commenced – all outstanding facets that give a new grower a little wiggle room!

Now let's look at every of these a little closer...

LETTUCE IN HYDROPONICS

Lettuce (and most different leafy greens) have to be your first plant to strive with a hydroponic system. These plants have a shallow root machine that fits their quick above-ground height. That potential there's no want to tie stakes or set courses for the plant. Instead, you just let them grow whilst oftentimes altering their nutrient solution. Eventually, they will seem to be appropriate adequate to eat, and you can!

Grow time: About 30 days

Best pH: 6.0 to 7.0

Tip: Stagger plantings so you have a non-stop grant of lunchtime lettuce!

Variety options: Romaine, Boston, Iceberg, Buttercrunch, Bibb

SPINACH IN HYDROPONICS

Spinach grows shortly in a hydroponic system, particularly when the use of the Nutrient Film Technique or different strategies that maintain the nutrient answer quite oxygenated. You'll additionally use far much less water than an in-the-ground garden. It's convenient to begin these plant life from seed and a week after sprouting, pass them into your system.

Grow time: About 40 days

Best pH: 6.0 to 7.5

Tip: For sweeter spinach, hold your develop temperatures between 65 stages F and 72 stages F. The decrease temperatures may additionally gradual grow time, though.

Variety options: Savoy, Bloomsdale, Smooth Leafed, Regiment, Catalina, Tyee, Red Cardinal

STRAWBERRIES IN HYDROPONICS

The worst component of strawberries is how seasonal they are. If you don't get them domestically when the crop is ready, you're relying on trucked-in berries that begin deteriorating as soon as they're picked. With agriculture, you'll have a ready-to-eat crop of strawberries all year long. Harvesting is super-convenient as properly – no bending over! Strawberries seem to do fantastic with an ebb and glide system, however deep water tradition or nutrient movie approach can do for a small crop.

Grow time: About 60 days

Best pH: 5.5 to 6.2

Tip: Don't purchase strawberry seeds, which won't be berry-ready for years. Instead, you prefer to buy cold-stored runners that are already at that stage.

Variety options: Brighton, Chandler, Douglass, Red Gauntlet, Tioga

BELL PEPPERS IN HYDROPONICS

Bell peppers are a slightly extra advanced hydroponic plant. Don't let them develop to their full height, instead, prune and pinch flowers at about 8 inches to spur pepper growth. Deepwater subculture or ebb and drift systems are best for peppers.

Grow time: About 90 days

Best pH: 6.0 to 6.5

Tip: Plan to provide up to 18 hours of mild for these plant life every day, and increase your light rack as the flora grow, keeping flowers about 6 inches from the lights.

Variety options: Ace, California Wonder, Vidi, Yolo Wonder

HERBS IN HYDROPONICS

There is a huge range of herbs that work wonderfully in hydroponic gardening. Studies have shown that hydroponic herbs are more flavorful and aromatic than those grown in the field. What herb do you choose to grow? Basil, chives, cilantro, dill, mint, oregano, parsley, rosemary, thyme, and watercress are all extraordinary options. Herb manufacturing is another gorgeous way to take a look at your new hydroponic system, and almost every gadget style is suitable for a spherical of herbs as you study the ropes!

Grow time: Varies via plant

Best pH: Varies via plant

Tip: Flush you are developing medium about once a week to get rid of any extra nutrients that your flora hasn't (or won't) absorb.

Variety options: Name your favorite, and you'll discover guidelines for developing it!

Part 2 - YOUR OWN HYDROPONICS

Chapter 8 - HOW TO START YOU OWN HYDROPONIC GARDEN

By definition, growing hydroponically potential a grower isn't the use of soil as a growing medium. The word hydroponics comes from Latin and potential working water. The intent and scope of this chapter is not to furnish a concise and whole set of instructions for developing hydroponically, however, to grant an ordinary overview of the quite several steps and approaches so beginner gardeners have a higher thinking of what may additionally be involved.

As always, do thorough lookup on each of the various elements yourself to maximize your hazard for success. All gardening can grant undesired consequences and trips will notably increase your chances.

Deepwater culture hydroponic developing uses no medium different than water for nourishing the roots. So, with this approach, keeping the water circulating and excellent aerated is critical to plant survival. Nutrient stages ought to additionally be held tightly at the acceptable tiers for plant vigor, and there is no real room for ignoring the basics. You can't just stroll away from this type of gadget for many days.

This method, although doubtlessly one of the best for plant yields, may also now not be for the novice. On the different hand, hydroponic systems that use a growing medium other than water have an awful lot extra room for error but represent a magnificent location to begin developing hydroponically.

Expanded clay, Rockwool, perlite, sand or gravel are popular develop mediums for hydroponic growing. The fundamental distinction between using these mediums and soil is that soil will generally maintain moisture lots longer than the others. Letting roots dry out in the course of the ordinary boom cycle is a sure way to lose them.

For this chapter, I'm going to use improved clay as an example. These are the little round balls of fire, naturally going on clay. They are pH neutral, reusable and don't compact. So, they maintain the identical moisture retention parameters over time. They do dry out fantastically fast, so you'll need to be careful when choosing your watering time.

Perlite has several similarities to clay pellets however as a result of several of the particles ar quite tiny (clay pellets ar astonishingly uniform in size), perlite can clog emitters. There are some other major concerns for a hydroponic system the usage of extended clay, which are:

There is no need to throw away the old media (as is from time to time wished with soil), so the long-term fee is reduced, and so is the hassle of getting rid of the ancient media)

It is lightweight, so it is less complicated to relocate large pots

Its numerous, tiny air pockets furnish a precise source of oxygen and assist maintain moisture

Preparation of the Grow Media

With improved clay (as with many other manufactured media), you'll prefer to wash and disinfect it earlier than use. You'll desire to do this originally and with every re-use. If it's not practical to use the grow pots for washing the pellets, find a massive container that without problems allows flushing and draining.

You'll want to be thorough to get all the previous roots and historic vitamins off when re-using the pellets. Initially, there is just some cleansing of dirt and precautionary disinfecting to do. Water is the first-rate for cleaning, simply keep washing. High-quality water is higher than attempting to use some cleaner that your vegetation won't respect later.

Disinfecting can be executed by diluted bleach or hydrogen peroxide. Dilute bleach to 10% and hydrogen peroxide to 3%. Disinfection prevents in opposition to microscopic pests in your garden, which don't certainly have many natural predators as soon as they infect the system, so making an effort to see there are as few as feasible before you begin will be properly rewarded.

Let the disinfectant stand on the media for a minimum of on hour. Make certain to wholly wash off any such disinfectant earlier than getting started. Mold can be an extra challenging pest to eliminate. Once it is detected, it is often critical to change the pellets.

Determining the Correct Watering Cycle for Your Plants

Is your device sluggish drip or fast? You will want a large pump and use more electricity for a quick drip. These emitters will normally feed for 5 to 10 mph (gallons per hour) each, and so do no longer have the funds for as desirable distribution of water as a sluggish 0.5- to 1.0-mph emitter.

Because increased clay pellets will dry out incredibly fast, do no longer have the watering set off for too long a duration of time. Avoid developing in areas uncovered to shifting dry air; this will motive the pellets to dry out prematurely.

Test your watering cycle out before obtaining started with plants. Popular watering cycles use on/off instances of 15 minutes and onwards. Try to keep away from an off length of extra than half an hour. If the use of a sluggish drip system, you can set lengthy watering times (say five hours or more) with 15- to 30-minute off intervals.

Slow drip pumps eat little power. The off cycles can prolong the existence of the pump. A 250-gph pump is successful in water up to 50 plants the place the drip charge totals 1 gph per plant. This price is not unusual. The timer you select will have a large bearing on your watering cycle.

You'll favor a water utility system that does a top job of distributing the water over the surface of the media. Only a drip or two is no longer going to do well. Ideally, you'll favor a drip for about for every 16 sq. in. or extra to minimize any dry zones in the media.

As the water travels down through the media, it will distribute itself outward via capillary action, so that as it travels lower, all of the media is utterly moist. Play with the timer putting to see that you get the satisfactory balance of moisture while keeping off intermittent drought conditions. Water will be captured in a reservoir beneath and then recycled over the media.

If using drip, use an in-line screen to remove particles from the water and forestall plugging of the emitters or getting older the pump. The pellets themselves will do quite a bit of elimination of particulates. Algae are naturally taking place in water especially where vitamins are involved. An in-line display or filter can help eliminate algae, however, you'll need to smooth it periodically.

Leaving vitamins in the water reservoir create algae and the algae limit the efficacy of your nutrients. Keep light out of the water tank retaining nutrients and clean it often.

Planting and Growing Tips for Hydroponics

Starting plants from seeds is a subject for many articles and describing how to do so would possibly take away the focus of simply searching at a single kind of hydroponic system. Probably the best technique I will be aware of is to use a Rockwool grow cube and incubate it in a humid environment. The water used for new roots needs to be as free from chlorine (or any different toxins) as possible. Using RO or distilled waterworks, or you can de-chlorinate the water yourself.

If you are transplanting from a soil pot, you will desire to gently dispose of the soil from around the roots and wash with first-rate water. Do not leave soil on the roots of a plant as this may cause microorganism or fungous infestation and plant disease.

Soil also will plug emitters in an exceedingly recirculated watering system. Make a cone-shaped moistened pellet vicinity where you lay the fully moistened roots and cowl them with wet clay pellets. Begin watering immediately. Avoid permitting new flora to have a dry cycle at all until they have installed for at least seven to 10 days.

When transplantation to the new pot, do thus as if planting clean root. Using a small net pot is convenient. Place pellets in the bottom, lay your root gadget over them and add pellets to the top of the root system.

When transplanted, the new root system is small and the water applied to the pellets needs to moisten the pellets the place the roots are. Using a net pot for these roots, and seeing that there are a couple of slow drips right in this area, will help.

If you are transplanting from a develop cube, area the cube neck-deep in the clay media in a net pot, and then transplant the internet pot into your full-sized clay pellet pot.

When immersing the small net pot into your develop pot, put you develop pot into a large bucket stuffed with water. The pellets will be in general suspended, making it a lot less complicated to insert the net pot into the media at the right depth, which will be base of the plant—do not submerse the plant! In this example, there are no soil particles to put off before transplanting.

Nutrition Tips for Hydroponics

When plant life is younger and currently transplanted, high-nitrogen fertilizer is going to stimulate the plant in the wrong way. Nitrogen is what produces foliar increase and new flora want to first develop greater roots. So, the nutrient method you'll want to use will exchange as your vegetation mature.

Often in soil, there already exists a huge quantity of macro and micronutrients, but in hydroponics, soilless gardening there generally aren't any except you add them. Again, this element is no longer rocket science, but it does require some understanding of what vegetation needs and what exceptional fertilizers to supply to produce healthy, vigorously developing plants.

Chapter 9 - HOW TO SELECT THE MOST APPROPRIATE GROWING MEDIUM

Choosing the Right Medium For Your Hydroponic System

In hydroponics, a "growing medium" is what you will use in the region of soil to region your plants' roots. When developing your hydroponic system, it is fundamental that you pick out a medium that fits your needs, will supply you the largest yields, and will be the easiest to maintain. Here are a few of the most popular developing media used in current hydroponic systems, and the blessings and negative aspects of each.

Rockwool – Rockwool is possibly the most popular developing medium used in contemporary hydroponic systems. It is a fabric made from basalt rock and which, which is melted and "spun" so the material turns into interconnected fibers. One of the primary advantages of Rockwool is that it retains water very well, which capacity that your vegetation are less probable to be harmed through dehydration if your pump fails. It also holds a remarkable deal of air, which capability that it will make it greater not going for your flora to be over watered. However, the dust and fibers from this growing medium can be hazardous, so you have to careful when you deal with it. Because this cloth has an excessive pH level, you might also want to pay exclusive interest to the pH stage of your nutrient answer to make sure the plants in your hydroponic machine stay healthy.

Coconut Fiber – Coconut fiber, on occasion known as coco coir, is, in reality, the powdered husks of coconuts. It is growing in recognition due to the fact it is one of the organic media available for hydroponic systems. It is recognized for its large oxygen and water capacity, which potential your plant life has a higher hazard of surviving if something goes wrong with your hydroponic system. However, some cheaper coconut fiber is acknowledged to incorporate massive amounts of sea salt, which can also harm your crop.

Perlite – Perlite could be a kind of volcanic rock. Perlite is one of the more low priced media you can find and is regularly blended with different media. Because of its price and good wicking action, this is the media normally considered in cheaper wick hydroponic systems. However, it doesn't hold water very well, and because it can be hazardous if ingested, you must additionally use a dust mask when handling.

Expanded Clay Pebbles – Clay pebbles are virtually created by way of baking clay in a kiln, developing a bunch of air-filled clay pellets. Expanded clay pellets are one of the extra pricey media you can use in your hydroponic system, but they may additionally clearly retailer you a bit of cash in the long run because unlike most different growing media, they are reusable. However, clay pebbles do no longer retain water or oxygen very well, and therefore may require you to mix them with some other medium to extend water retention.

Air – Using air as a medium, which is also occasionally referred to as using "no medium," is very cost-friendly, because it if truth be told capacity that you don't have to buy a medium. Since your roots are constantly uncovered to air, you can additionally continually be guaranteed that they are continually getting the oxygen that they need. Bear in mind, however, that using air as the medium in your hydroponic device leaves little room for error. If your pump fails, your roots can dry out in a count number of minutes, significantly detrimental or even quickly killing your complete crop.

Chapter 10 -
CHOOSING PLANTS

What Can You Grow Hydroponically?

People ask frequently, "what can I grow hydroponically? The reply is in reality quite simple: You can grow a large range of flowers, vegetables, and herbs hydroponically, except for mushrooms that are fungi.

Following is a checklist of many vegetation that develops nicely in hydroponic systems, together with some statistics of interest:

Flowers

Growing flowers lends itself fantastically to hydroponic gardening as they can be grown in large numbers, and can be grown year-round. Most flowers will do properly in a hydroponic garden, and when seedlings are big enough, plant life can be reduced or transplanted.

Herbs

Many herbs will develop very well in a hydroponic setting. Some that do the satisfactory encompass anise, basil, catnip, chamomile, chervil, chives, cilantro, coriander, dill, fennel, lavender, marjoram, mint, oregano, parsley, rosemary, sage, tarragon, and thyme.

Anise

Anise is a feathery annual that grows from 1 to 2 ft high, has finely cut serrated leaves and very small, whitish vegetation in flat clusters. Both the leaves and seeds have a warm, candy licorice taste. It grows swiftly from seed and need to be planted after all threat of frost has passed. The inexperienced leaves can be reduced each time vegetation is giant enough and seeds may be gathered 1 month after plant life bloom. Anise leaves can be used in salads and as a garnish; the seeds flavor confections such as desserts and cookies.

Basil

In a blanketed environment, developing basil can be carried out during the year. Once mature, it can be harvested and trimmed weekly. It responds extremely properly to hydroponic growing.

Cannabis

A mind-altering herb derived from the flowering topnotch of hemp plants. Cannabis is controlled underneath Schedule I of the Controlled Substances Act of 1970. It is additionally known as bhang, ganja, grass, hashish, marijuana, pot, reefer, tea, and weed. It prospers and grows to an extra energetic plant in a hydroponic system. Most cannabis plants cultivated in the United States commence to flower by way of late August to early October and the flowers are harvested from October to November.

Rosemary

A hardy evergreen sub-shrub grown basically for its fragrant leaves that are used in culinary seasoning and yield an oil once used in medicine. Small mild blue plant life are borne in April or May. The foliage is white and woolly on the underneath facet and dark and vivid above. Plants can develop to a top of 6 toes and ultimate for years but want safety from the cold. It prefers alkalic soil and full sun, but does tolerate moderate shade. Sow in seed apartments 22 weeks before sale in 10 cm diameter pots. Seeds to finished plugs, 12 weeks; plugs to saleable plants, 10 weeks.

Sage

Common title for the hardy sub-shrub that is considerably grown for seasoning dressings used with wealthy meats, and for flavoring sausages and cheese. In hydroponics, it can be grown from seeds covered from cold and it prefers full sun. As the vegetation often exceed 3 ft in diameter, they have to be grown at least that some distance apart. Sage leaves should be harvested before blooming and dried in a well-ventilated room on monitors or in a business dryer, away from direct sunlight and then shop in airtight containers. Sow in plugs or seed flats 12 to 14 weeks earlier than sale. Seeds to completed plugs, 8 weeks; plugs to saleable plants, four to 6 weeks.

Tarragon

A perennial herb the leaves of which are used for seasoning, in particular, vinegar. Tarragon grows to two or three ft tall and likes moderate sun, preferring some color throughout the most up to date section of the day. Tarragon, at some stage in growth, appears to have little aroma; but after the leaves or tops are harvested, the oils listen and begin emitting their special tarragon candy smell. Plugs to saleable plants, 7 weeks.

Thyme

A plant of the mint household long cultivated and valued as a candy herb. It has small lavender or purple vegetation and is grown as a border plant, for ornament, or as an herb to be used for seasoning. Thyme should be planted in early spring. It is terribly hardy and can grow underneath most conditions. It prefers full sun. Thyme needs marginal fertilization once full-grown in a very agriculture system. Sow in plugs 12 to 14 weeks earlier than sale. Seeds to completed plugs, 6 to eight weeks; plugs to saleable plants, four to 6 weeks.

Watercress

Low developing and trailing European perennial, a member of the mustard family. It is without problems grown from seed. Its herbal season is from mid-autumn until spring. After its flower buds appear the leaves become too rank in taste to be edible. It is additionally effortlessly grown indoors in a hydroponic system. Start flora with seed by using sowing gently in pots stuffed with a medium. Watercress has many culinary, decorative, and medicinal uses.

Vegetables

Vegetables that had best in a very agriculture garden include artichokes, beans, lettuce, spinach, cabbage, beets, asparagus, broccoli, cauliflower, Brussels sprouts, and peas. Vegetables that grow at a lower place the soil, like onions, leeks, carrots, parsnips, potatoes, yams and radishes will grow hydroponically, however, they will to boot need larger care. Some vegetation to keep away from are corn, zucchini, summertime squash, and vining plants. They can be grown in a hydroponic garden, but they are now not housed efficiently, and simply no longer practical. They will dominate your whole unit. Your sources are higher spent on crops greater ideal to the compact systems.

Chapter 11 - TRANSPLANTING

How to Clone Hydroponic Plants and Transplant

Cloning your plants, or what is commonly referred to as cuttings, is a splendid way to take a cutting off a plant and to then be in a position to grow the cutting. Done properly, you may want to take this cutting and develop an entirely mature copy of the plant it used to be taken from. This is a tremendous technique to use when growing hydroponically permitting you to take a small batch of one of your pinnacle high-quality plants and clone it which, in turn, ought to grow to double or extra than what you began with.

Taking cuttings from plant life can be a little complicated when first trying it, and it takes very distinctive care to get the clone up and going once cut.

When hydroponic farming, making clones ought to be the key to your success, and here we will supply you some prevalent hints on clones and beyond.

The first thing you do is some lookup and find out where the nice region is to take a reducing for cloning for your precise plant. Different plant life clone satisfactory from exclusive places, however, cuttings are typically usually taken close to the plant nodes or branches.

1. With a thoroughly cleaned (or new) razor blade or even very sharp scissors, make a reduce at an angle.

2. Then rapidly region the reducing in a starter developing medium such as Rockwool cubes.

3. Put the clone in a humidity dome, or germination box. At this point, dont put it at once into your hydroponic device as there are no roots to be fed and accordingly will die notably quickly. The aim here is to get the surrounding environment for the clones very high in humidity as this moisture in the air will be their only supply of water for a few days or so.

4. The subsequent seven days will be critical and the reducing need to be kept correct moist at all instances to stay wholesome till the root develops.

5. After seven to ten days, you ought to begin to see root improvement and increased growth – this is a good sign.

6. Keep them where they are for any other three to 4 days till there is prolific root growth coming from the bottom.

7. Transplant into your hydroponic farming system.

8. Carefully area the clone into the growing medium used in your hydroponic system, if the use of Rockwool, this can without difficulty be transplanted into any system.

9. Once you acquired all your clones in place, it is a correct idea to mist them with water a few instances a day, just for a few greater days.

Here are just a few things to maintain in thinking when taking clones:

Always use a sterile blade or scissors – if a soiled blade is used, the chances for survival are reduced drastically as the open wound on the plant will enable sickness to spread.

From the time you take the slicing to approximately two weeks later (i.e. 10 to 13 days), it is very necessary to keep the cuttings moist at all times. If the clone is allowed to get too dry, it will rapidly begin to die as its only supply of water and vitamins is the quantity you give it.

When transplanting to a hydroponic farming system, it usually makes positive the roots are huge ample for them to properly acquire the nutrient answer in the system. If the roots are too short, then wait an additional few days for the roots to advance more.

Chapter 12 - HOW TO SET UP YOUR OWN HYPROPONIC GARDEN

5 Easy DIY Hydroponic Plans You Can Build in Your Garden This Weekend

You don't want a massive garden to develop your sparkling produce. Nor do you need years of journey to construct your personal DIY indoor develop system. That is the beauty of hydroponics.

The complete discipline is based on flexibility and inventiveness. There are ratings of DIY hydroponics plans floating around the World Wide Web.

Here is a decision of the nice self-made hydroponics plans absolutely everyone can build. These plans consist of beginner, intermediate, and professional stage setups.

1. The Passive Bucket Kratky Method

The Kratky Method is no doubt one of the easiest hydroponic plans you can begin through yourself inside numerous hours.

This machine is terrific for all and sundry who just receives began with hydroponics. What you need is a bucket, some developing media (like hydroton, perlite), some internet pots, hydroponic nutrients, and pH kits. These are all required to set up a passive device (no electricity required) that can run robotically for weeks barring maintenance.

You can develop inexperienced vegs like lettuces, spinaches at the start or fruit flora like tomatoes after you have bought enough experiences.

Difficulty level Beginner (1 / 5)

2. Simple Bucket Hydroponic System

This are some other easy hydroponic setups for beginners. All you want is a 5-gallon bucket, some growing media like coco coir or perlite-vermiculite, and nutrient mix.

The setup works by using the use of the developing media to make a capillary action, which moves nutrients up to the flower roots.

This system is perfect for single large plants. If you want to preserve things basic, you can water the machine manually.

For an automated system, you will need every other bucket for the reservoir, and a submersible pump, and a timer.

Difficulty level Beginner (1 / 5)

3. Simple Drip System With Buckets

Another entry-level option, this is a bit extra advanced than the single bucket gadget above. It can nonetheless be cobbled together using components that value much less $100 in total.

The original graph calls for developing four plant life in separate buckets, all fed utilizing a frequent reservoir. This is a very flexible setup that can be improved in the future.

You can trade the size of the containers, and reservoir relying on the size of vegetation involved. You can use large 4-gallon buckets or smaller containers.

Remember to buy a larger reservoir in case you want to add more vegetation to the mix later on.

Difficulty level Beginner (2/5)

4. Aquarium Hydroponics Raft

This is a very cool challenge to get your toes wet in the world of hydroponics. It is additionally an awesome way to get your youngsters hooked to the field.

As the title suggests, you will want an aquarium fish tank to make this work. This device can be used to develop small beans or even a single large lettuce.

Along with the regular ingredients like nutrients, water, and plants, you will want a raft of barge long-established out of foam. The gadget can be passive or active, using pumps and electricity.

Difficulty level Beginner (1.5/5)

5. PVC NFT Hydroponics System

Large four-inch PVC pipes can be used to create your self-made hydroponics system. In this plan, the plants are positioned in cups that are arranged in holders drilled into the pipes.

The system is watered using a reservoir and pump. This is a closed system, with the water circulating between the pipes and the reservoir.

This plan is best for growing a lot of small flowers in a small area. The fundamental gadget can house anywhere from 20-40 plants.

This gadget can be positioned indoors or outdoors. If indoors, develop lights are of course essential.

The hydroponics technique used in this plant is referred to as NFT. It is an excellent design for developing flowers like tomatoes.

Difficulty level Advanced (4/5)

Chapter 13 - PROBLEMS AND TROUBLES

8 Common Problems With Hydroponics

(And How To Fix Them)

Hydroponics is a remarkable way to grow flowers at home that is challenging, fun and very rewarding. However, there are a variety of problems with hydroponics that you might also encounter, and it is necessary to examine to avoid these or deal with them successfully.

Hydroponic developing is a greater technical ability than a growing flora in soil. You can examine a lot from reading books and articles, and gazing at instructional videos. However, one of the best approaches to research is from our mistakes.

1. Hydroponics System Leaks

System leaks can appear for an entire range of reasons. Leaks can happen at any joins or valves in your system. They can additionally manifest if your machine gets blocked, such as when the root mass clogs up an NFT system, main to water backing up and overflowing. Leaks can additionally take place if you construct a machine with a reservoir which cannot maintain all of the nutrient solution in the system. In this situation, electricity reduces or pump failure, which might also lead to back up and overflow of your reservoir.

Solution

Test your machine before planting anything. Tighten any valves and make sure all connections are tight and secure.

Regularly take a look at your system for problems such as root overgrowth or clogged drains or outlets.

Ensure that you pick out a reservoir which can comfortably keep all of the nutrient answer in the system, not simply the quantity that is in it when the device is in use.

If you are the use of an indoor system, reflect on consideration on setting it on a water-resistant floor or, if possible, on a drip tray if you are using a small system. This is an excellent thought to seize leaks, but will also minimize mess when tending to your system.

2. Buying Cheap, Insufficient Or Incorrect Lighting

I like to use my hydroponics structures indoors so that I can grow clean greens all 12 months-round. Without ample lights of the right type, the performance of a device will be very disappointing.

I've made more than a few errors with indoor grow lights, such as buying lower-priced lights that had been inadequate for what I needed or buying the wrong type of lights that led to negative fruit and vegetable yields.

Solution

For most people, I would strongly advise looking at LED and T5 fluorescent develop lights. These are usually the easiest to use and will be appropriate for most users.

If you are buying LED develop lights, do no longer go for the most inexpensive option. Do a bit of research and purchase first-rate lights that will produce mild at the right wavelengths and in ample portions for your system.

Ensure you buy ample grow lighting fixtures for your system. A true rule of thumb is to calculate the square photos of the cover of your grow vicinity and multiply this by way of sixty-five

Here is a speedy example.

A developing place of 4ft by 6ft. Total vicinity = 24 sqft.

24sqft x 65 = 1560 watts

For this developing area, you will want approximately 1560 watts of grow lighting. This is an excellent rule of thumb, and is what I commonly stick to.

3. Using The Wrong Fertilizer

When growing flora in soil, many of the micronutrients wanted are already current in the soil in sufficient quantities. For this reason, fertilizer designed for growing plant life in soil does no longer wants to encompass many of the trace micro vitamins that are crucial for healthy plant growth.

Solution

Make positive you buy vitamins designed for use with hydroponics.

You can make your very own hydroponics fertilizer from scratch, but it is a good deal easier to purchase a two or three-section solution. This can be mixed to produce nutrient answer that can be adjusted to most plants and increase phases.

4. Not Keeping Things Clean

If you let your hydroponics setup and the location around it become messy and dirty, you may additionally extend the chance of spreading sickness or pests to your hydroponic system.

Part of the cleansing technique is to cease algae, illnesses, and pests from being in a position to set up themselves in your system. Whilst some people do run structures particularly designed to encourage the growth of really helpful bacteria, I think for most home hydroponics setups, it is better to avoid the pathogenic organisms, by commonly cleaning your machine and surrounding area.

Solution

Keep the place round your hydroponics setup clean and well organized.

Every 2-3 weeks, drain the system, flush the growing media and roots with water and smooth the reservoir, pumps, and tubing.

5. Not Learning As You Go

Every crop of vegetation in a hydroponics machine is different. Some matters will go nicely and you will encounter some problems, both minor or major. You take the opportunity to analyze what went properly and what went wrong, to regulate your exercise for future crops.

Solution

Document, picture and take the word of the good and awful factors of every machine you use and crop you grow.

When you come upon a problem, look for a solution. Books, websites, and Youtube have so plenty of facts reachable that you will be able to solve your troubles or prevent them the subsequent time.

6. Not Monitoring The Health Of Your Plants

If you do no longer monitor your plants frequently, you will pass over the early signs and symptoms of problems. Whether this is an insufficient increase or symptoms of deficiency or disease, the earlier you recognize there is a problem, the extra chance you have of correcting it and not ruining your plants.

Solution

Monitor the boom and circumstance of your flora frequently.

When you see a problem, take the time to find out what the trouble is and strive to correct it.

If you be aware disorder or pests, deal with

early and you may also be in a position to stop excessive injury to your plants.

7. Not Monitoring And Adjusting the pH Level

The pH degree of your nutrient answer is one of the most fundamental aspects of hydroponic growing. When growing plants in soil, the soil itself acts as a pH buffer and prevents fast modifications in the pH level. This ability that pH issues are slower to enhance and can be dealt with more easily.

This is no longer the case for hydroponics. The pH can trade significantly over hours or days due to a variety of factors consisting of temperature, rate of absorption of nutrients by your plants, presence of disease, excess evaporation, etc.

Solution

When developing with hydroponics, you must monitor the pH of your nutrient solution.

In a new system or when current modifications have been made, you can also need to take a look at and adjust the pH daily. In a steady system, you can decrease testing to once or twice per week. As you attain a journey with hydroponic growing, you will begin to apprehend the factors that can have an impact on the pH and you will get a feel for how frequently to test.

The fine preferences for checking out pH are to use a pH trying out kit or a pH testing meter. I commonly propose getting a first-rate high-quality electric powered pH trying out meter, as it makes pH checking out quick and easy.

8. Nutrient Deficiency and Toxicity

Several elements can motive nutrient deficiency or toxicity in your plants. It's no longer usually handy to tell which nutrient is inflicting the problem or whether or not deficiency or toxicity is the problem. There are several signs to look out for to become aware of deficiency and toxicity of many nutrients, and you will get better at identifying issues with time and experience.

pH, temperature, plant boom rate, nutrient answer concentration, consumer error and a total host of different elements can purpose nutrient problems. Don't forget that extra tiers of one nutrient can purpose problems with the absorption of another.

Solution

Make sure to make up your nutrient answer cautiously and accurately.

Ensure that the water you are the use of to make up your nutrient answer is no longer excessively hard. If so, think about diluting it with distilled water, or using water that has been through a reverse osmosis filter or activated carbon filter to limit the degree of dissolved solids.

Monitor the concentration of your nutrient answer with a PPM/EC meter

You May Also Like: How Do You Care For A Guzmania Plant

Monitor and regulate the pH of your nutrient solution.

If your plant life starts to display signs of nutrient deficiency or toxicity, my recommendation is to flush your system, discard the nutrient answer and make up a clean batch. More skilled growers may additionally have the skills to modify matters as they go, but most beginners and intermediates will be higher to take the protected approach.

Chapter 14 - COMMON MISTAKE TO AVOID

As you commence to navigate the world of hydroponics, examine from these errors and preserve them in idea when beginning or scaling your personal system. Doing so will shop you a lot of heartache and perchance monetary despair associated with these seven mistakes.

Mistake #1: Growers layout unusable or hard-to-use farms

Designing an unusable farm is a mistake of inexperience greater than whatever else. Many growers haven't grown before (at least now not on a large scale), so they don't think about factors like workflow and efficiency. This results in farms that:

- Don't use house efficiently
- Are difficult to harvest
- Require lots of transplanting and tending
- Aren't conducive to pest control

- Don't allow effortless get entry to to necessary components

Since labor is regularly the largest variable cost on farms, labor-efficient format is important. The therapy for this mistake is to suppose carefully from the begin about how you will use your system.

Consider all of your variables, from growing desires (light, water, nutrients, pests) to person desires (access, convenience, automation, redundancy) from the beginning, and solely begin to format your machine when you've critically considered these variables.

Talking to mounted growers and travelling their system designs can be a super help as well. Be certain to ask questions and locate out what they would do in a different way if designing their structures today.

Mistake #2: Growers underestimate manufacturing and device costs

Most growers beginning out in farming fail to completely understand their costs. They get started, invest in large facilities, pricey utilities, and equipment, however in no way get the threat to thoroughly make use of them because the price range is ate up via unanticipated costs. Some generally forgotten costs are:

- Packaging
- Pest controls
- Insurance
- Labor
- Printed marketing materials
- Ongoing maintenance
- Heat removal
- Equipment replacement

These are foremost fees that add up. The cardinal sin is that most commencing growers vastly underestimate the price of labor—whether it's their personal or anybody they've hired. Raft manufacturing is an instance of a common, labor-intensive, hydroponic manufacturing technique.

For raft systems, the labor expenses can be significant—as a whole lot as 45-60 per cent of whole costs. Most producers don't even consider this in their labor estimates, so when the fee of harvesting and processing comes in, the bottom line quickly drops from the black to the red.

Mistake #3: Producers confuse biological viability with monetary viability

There is a false impression that beginning a farm enterprise is ninety per cent developing and 10 per cent selling. In our experience, it's simply about the opposite. Most farmers make one of two errors in this regard.

First, they don't account for the time and monetary charges of getting their produce to market once it's grown, and, as a result, they don't finances sufficient time or money to correctly promote their produce.

Second, they plan the biological feature of their farm (technique, crops, equipment) except testing the feasibility in opposition to their markets. Does local demand suit their specific kind of produce?

If not, then they're caught with a facility and a lot of produce but no one to buy it. The bottom line is that it doesn't be counted how healthful your plants are if you don't have the ability to promote them.

Mistake #4: Growers pick out the wrong plants for their local weather or technique

It's handy to be seduced by way of flowery descriptions of unique new crops that populate so many seed catalogs these days. If I had each dollar returned that I've wasted over the years trying to develop vegetation that is either: a) no longer applicable to my production approach or local weather or b) not in demand in my neighborhood markets, I'd have a hefty chunk of exchange back from seed companies. Before you pick outcrops, you need to ask a few questions:

- What constraints are placed on growing utilizing your climate?
- What growing technique will you be using?
- Can you grow this crop with your manufacturing technique?

Different vegetation have one of a kind needs, and some can solely be cultured in sure ways. Folks using rafts have to no longer be attempting to grow tomatoes. Similarly, people with the usage of beaten granite media have to now not expect to be capable to produce marketable root crops.

If you live in the Northern Hemisphere, trying to develop lengthy day-length crops in an eight-hour day won't work nicely for you. If you're in the south, and constantly fighting the heat, tries to develop a cool-weather crop like rhubarb would be a terrible decision.

Be thoughtful about what you grow.

Mistake #5: Growers choose the incorrect market

Another component that has to be regarded is your market. Whether you're growing for your household or a farmers' market, you're nevertheless immediately or in a roundabout way promoting your produce. Growing a crop that no one needs is a waste of your time and money. When I was developing up and zucchini season hit, all and sundry used to be attempting to unload zucchini on unsuspecting neighbors and friends.

The domestic gardeners in our region made the mistake of growing an easy-to-culture, but unwanted, crop. There's only so an awful lot zucchini a human can consume. Analyze your market carefully. Consider what your opponents are growing.

If you live in an area the place summer season opposition is fierce from subject producers, then pay attention to something they can't develop throughout that period. Most likely, if a restaurant client wants neighborhood natural lettuce and a discipline producer will sell it at 50 cents a pound, you won't be capable to keep that consumer over the summer.

Figure out what you can do to make ends meet in mild of this seasonal competition, or lock your clients into long-term buying contracts. The backside line: select a crop with a guaranteed market.

Mistake #6: Growers operate structures that have terrible tune records, then count on distinctive results

When you're questioning about implementing a system, don't be sold on the supposed profitability. Ask for references for gadget customers that have been in commercial enterprise for quite a few years. If they can't supply them, walk away. Interview references cautiously to find out whether or not they're profitable and doing well.

For example: Raft designs can be very productive and profitable in areas where greenhouse manufacturing is no longer required for most of the yr and where labor is fairly inexpensive. In northern climates, however, greenhouse raft production is without a doubt now not price effective, as evidenced utilizing the lack of hooked up business raft growers in the northern United States.

Although many are drawn to raft production because of the low start-up costs, the terrible productiveness per square foot of greenhouse area capacity that pricey sources are no longer used as correctly as they ought to to be a viable business.

Mistake #7: Growers grow too big, too fast

Going too huge too quick is a common mistake. This leads many establishing growers to get funding for large, luxurious services before they apprehend their cost structure or the market they're attempting to service.

Growers that grow too quick also seem to have catastrophic failures extra often.

Big system failure capacity massive cash failure; more importantly, system failure causes a hole in furnish to clients who want constant delivery. When this happens, these clients begin to look elsewhere, and by way of the time the grower is returned online, he's regularly misplaced many valuable clients.

These are screw-ups that threaten the complete enterprise. Growing slowly, on the other hand, requires patience, but approves growers to grow into their market organically, assembly nearby wants and needs with products. Large entrants tend to flood the market with products that they consider are desired—often with combined results. There are three things that you can do to keep away from the pains of growing too fast:

- Rein in the desire to overwhelm the market.
- Develop a niche market.
- Get creative and supply value
- Traversing the mastering curve with grace

Every farmer, whether professional or green, experiences a gaining knowledge of curve when they start constructing out a new system. This much is inevitable. However, gaining knowledge of the curve doesn't have to characterize losses and pain. Smart planning is the fantastic issue you can do for your farm, for although novice errors are inevitable, large losses don't have to be.

Hundreds of tools—from farm planning software programs like Able to learning applications like Upstart University—exist to help new growers and new commercial enterprise owners. Take benefit of them as a lot as you can.

While developing hydroponic plants, most of the growers commit the same frequent mistakes which wreck their whole garden. With ideal research and planning, you can avoid several such common mistakes. One of the biggest errors growers typically commit is that they start developing hydroponics plants besides having any simple knowledge about how to develop and look after hydroponically grown plants. Before you start your hydroponic garden, check out our story that offers you confidence in hydroponic growing!

One frequent mistake most of the gardeners commit is to now not provide enough air motion internal their hydroponic garden. Air movement is imperative to plants' breathing as it provides clean air to your leaf zone. The air in your grow room ought to incorporate enough oxygen, CO_2, however, it has to now not include molecules of industrial pollutants, particulates, and other airborne debris. Proper airflow is quintessential to get higher yields in hydroponic gardening however you simply can't throw a fan in your hydroponic develop room. You need to put a proper measurement fan as hydroponics fans are an essential section of your indoor backyard set up and assist to manage airflow, heat, and other environmental conditions. If you are the usage of HID grow lights, ventilation becomes even more necessary due to the fact of the amount of warmness these lights produce. Always be mindful to preserve the fan inconsistent action and it needs to not blow at once on the vegetation as this can motive dehydration.

Hydroponics Gardening

Hydroponics flowers thrives in an exceedingly sleek and well-maintained surroundings, thus you must clear all the junk like fallen leaves, dirt and alternative materials that may attract and breed diseases. You need to hold your develop room dry to keep away from any sort of fungal infestation. Do not smoke, eat, or allow pets close to the plants as these can grant damage to your plants.

To avoid these frequent errors in your hydroponic garden, you want to pay shut interest to your plant life each day to make sure that your garden is walking smoothly. Every time you visit your garden, be mindful to test your pump systems, reservoirs, water levels, pH, nutrients, lights, timers and plants.

Chapter 15 - TIPS AND TRICKS

1. Maintain a steady temperature and improve climate control with effortless develop reflective sheeting.

2. Ensure suited nutrient mixing and oxygenation of your reservoir by the use of an air bubbler and air stone.

3. Maintain pH tiers with a Prosystem Aqua computerized pH dosing pump, and avoid essential nutrients being locked out of your plants.

4. Ensure your dark durations obtain no light by using using lightite sheeting for a complete black out.

5. Maximise the availability of light, minimize power prices and cast off hot spots using diamond lightite sheeting.

6. Replace your growing bulbs regularly. Lights lose lumens shortly and can degrade through as plenty as 30% inside a year.

7. Add CO_2 to your grow room the herbal way the usage of Exhale CO_2 baggage and let your plats breathe.

8. Avoid the degradation of your treasured nutrients by means of storing in a darkish location and heading off publicity to sunlight. Green Planet vitamins are all saved in black bottles to avoid this happening.

9. Enhance the biology of your plant using Myco Fusion mycorrhizae powders.

10. Improve nutrient uptake via the usage of Fulvic and Humic Acids.

Conclusion

The bodily hydroponic developing surroundings are akin to the idea of greenhouse gardening however on a whole lot large scale with automatic rotation of the beds for maximum light exposure at all times, as nicely as the loading and unloading of vertical developing beds in incredibly managed developing environments. The common greenhouse, also recognized as a hothouse or glasshouse, targeted solar heating while ventilation was once manually operated by way of opening and closing window panels, and lighting was once herbal and diurnal.

The planting medium used to be soil so the traditional plant-soil issues have been present and water was once supplied regularly in overabundance. Humidity and condensation have been ongoing problems that wanted to be addressed but were no longer managed in a greenhouse as such. The conventional fashion of the greenhouse is now supplanted by way of the vertical A-frame developing systems. Lighting is no longer completely based on natural mild sources however uses LED and sulpha plasma lights systems.

The challenges are nonetheless existing and very real. The largest task is associated with the ambient temperatures surrounding the greenhouse. Controlling the inner temperatures of the greenhouses has required complex and superior airflow and cooling techniques. The science of the entire business enterprise is no longer simply a local initiative but has required global cooperation and partnering with information from the U.K., Japan, the Netherlands and many different components of the globe.

In addition to new and extended science and modern agricultural practices, the stage of staffing is also modernized with surprisingly educated and conscientious people who recognize the production device and the challenges of working in a laboratory fashion environment.

www.ingramcontent.com/pod-product-compliance
Lightning Source LLC
Chambersburg PA
CBHW060841220526
45466CB00003B/1194